COMMUNICATION and INTERPRETATION

THE LEADING EDGE was created in 1982 as a medium of communication between members of the geophysical profession worldwide. It was designed to complement GEOPHYSICS, our archival journal of geophysical technology. SEG's Executive Committee, in its desire to improve the communication of *TLE*, created the Editorial Board in early 1985. The Board has the responsibility to set publication guidelines, to solicit articles, to review contentious material and to organize special publication projects.

John Hyden told us in Notes from Headquarters in February 1986 that 53 percent of SEG's membership consider themselves interpreters and that 70 percent of the membership work in the seismic field. It follows that seismic interpreters occupy close to half of the total membership. Several of them have spoken out in Signals complaining that SEG's journals do not provide the type of material they want to read. The Editorial Board completely understands this complaint but also agrees with Frank Levin whose letter in Signals in November 1986 reminds us that "Editors can only publish those papers that are submitted to the journals they edit."

Thematic guidelines can provide a framework within which material can be gathered and authors motivated. The Special Issue concept seems to have been well received. Several series of articles are now being formulated, each series having a theme. The hope is that a series, once established, will provide an expectation of interesting material for the readership and a snowball effect to generate more articles from willing authors.

The first series concerns Seismic Interpretation and is technique oriented. The Seismic Interpretation series will, if it receives readership approval, continue indefinitely with articles appearing every one or two months...

The above comments have been excerpted from the introduction I wrote to accompany the first article in the Seismic Interpretation Series in THE LEADING EDGE. That article was published in March 1987 and others have appeared approximately every two months ever since then and we expect that this regularity will continue for several years because a snowball effect is definitely starting. We are very pleased with the unsolicited response, but it is not yet overwhelming. Please let any member of the *TLE* Editorial Board or staff know if you have something to contribute or know of someone who could. Future solicitations will be aimed at filling gaps in the coverage of the series to date. We hope that ultimately the series will be a rather comprehensive review of seismic interpretation techniques in use today.

This volume collects the first 12 articles in the Seismic Interpretation Series in a format that should make them readily available for consultation. Subsequent volumes are expected to be published at yearly intervals. We hope that this formula satisfies the demand and provides a useful, cohesive, and archival compilation for the world's seismic interpreters.

ALISTAIR R. BROWN
Past-Chairman, TLE Editorial Board
January 1989

Seismic interpretation in salt-controlled basins

By JAMES FOX
Phillips Petroleum
Bartlesville, Oklahoma

(Author's Note: Two years ago, I wrote a letter to the SEG, which was subsequently published in THE LEADING EDGE, *decrying the lack of applicability of most SEG material to the seismic interpreter. A. M. Olander, then President, answered my complaints by stating that a fundamental change in the SEG would only come about from an active movement by its members. The last two years have seen a great change in the Society to better meet the needs of the seismic interpreter, and I wish to commend all those involved in the process.)*

Salt-controlled basins are common worldwide, and many of them hold major reserves of hydrocarbons. Examples of producing salt-controlled basins include the Gulf of Mexico, the North Sea, the Persian Gulf salt basins, and the South Atlantic salt basins offshore Brazil and West Africa. Because of the major hydrocarbon potential of these basins, many exploration geophysicists are currently working there . . . or will work there in the near future. Also, because many are in a mature exploration stage, explorationists must focus their investigations on the subtle trap. Searching for subtle traps in salt-controlled basins can only be accomplished through understanding the models which govern the effects of salt tectonics on the surrounding geology.

This paper will attempt to provide coverage of two main points:

• A brief overview of salt models and the interaction of salt growth and sedimentation.

• An interpretation procedure designed specifically to search for the subtle trap in salt-controlled basins.

Salt growth models. Salt deforms in a predictable series of stages (Figure 1). These stages are controlled by various factors, including:

• Configuration of the presalt section

STAGE 1 – DEVELOPMENT OF SALT SWELLS AND WALLS

Salt movement is lateral, not vertical. Large, linear salt walls form with large peripheral sinks (depotroughs) between them. These walls may localize along discontinuities in the presalt surface, such as faults or folds.

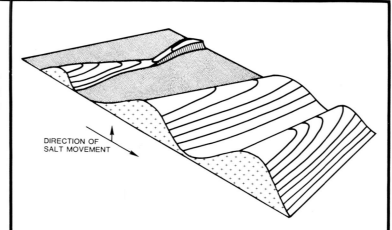

STAGE 2 – LOCALIZATION OF SALT INTO MASSIFS

Salt movement is still predominately lateral, but perpendicular to the direction of movement during stage one. The salt walls localize into massifs, but remain non-diapiric. Saddles between salt massifs allow sediments to pass across older salt walls. Turtle structures form during this stage, as do basin edge faults.

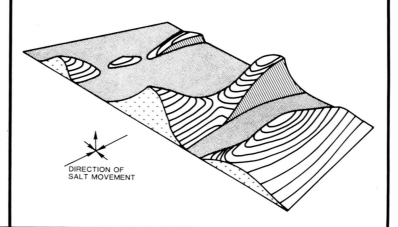

STAGE 3 – FORMATION OF DIAPIRS

Salt movement becomes vertical during this stage, and the salt bodies become diapiric. Extensive faulting of the surrounding section occurs, and significant erosion or nondeposition occurs across the crest of the sediments overlying the diapirs. Diapirs assume various shapes, depending upon external factors and internal salt geometry.

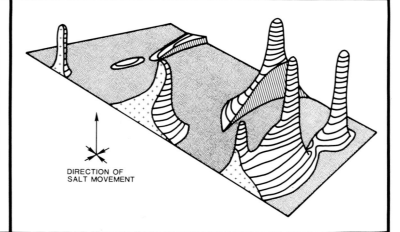

Figure 1.

- Amount of subsidence
- Rate of sedimentation of the overburden
- Thickness of the original salt
- Basinal position

Depending on such factors as basin type, and the balance between subsidence and sedimentation, some of these salt controlling factors are more important than others.

Regardless of which of these controlling factors dominates in a given basin, salt still deforms in a series of predictable stages. Salt is deposited as a flat lying bed, commonly on an irregular surface, resulting in differential thickness. This initial deposition usually sets up two major zones of salt formation which deform differently: the basin margins where the original salt was fairly thin and has a simple deformation history, and the basin centers where the original salt was thick and deformation is much more complex.

Salt becomes mobilized after a sufficient thickness of overlying sediment is deposited. As sediments accumulate on the basin margins, the sediment load causes the salt to flow laterally. The movement is gradual, forming large salt walls parallel to the basin margin. The salt will continue to move laterally until it reaches some discontinuity in the presalt surface, usually developed by presalt faults related to rifting or initial opening of the basin.

When this occurs, the salt will localize into salt walls or ridges. During this stage of salt movement, noticeable thinning occurs in the sediment overlying the salt walls, whereas thickening occurs between the salt walls. These zones are called depotroughs or primary peripheral sinks.

As sedimentation increases, the second stage of salt growth is initiated. The salt walls localize into separate structures called massifs or nonpiercement salt domes. The salt still moves laterally, but the main direction of movement is perpendicular to the previous direction. As salt withdraws from the adjacent area into the salt massifs, several salt withdrawal features form. These include:

- Secondary peripheral sinks
- Turtle structure anticlines
- Counter regional faults
- Secondary depotroughs or depopods

These features form because the deposited sediments must adjust and fill in the voids left by the departing salt.

The third stage of salt movement occurs when the salt becomes diapiric, piercing the overlying sediments and forming salt diapirs and stocks. This period of movement is characterized by extensive faulting in the sediment section surrounding the diapir.

Faults related to salt tectonics. Faults abound in salt-related basins. To properly

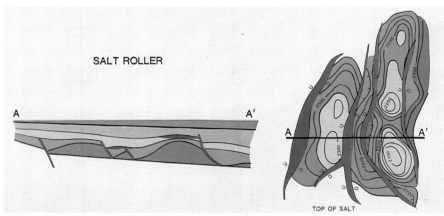

Figure 2.

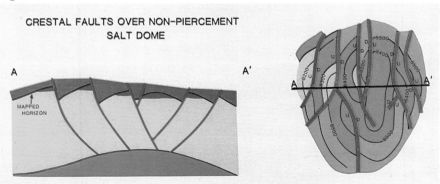

Figure 3.

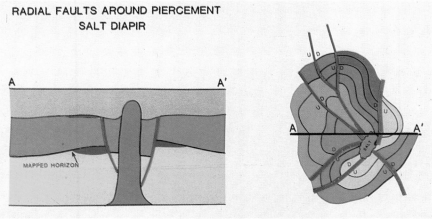

Figure 4.

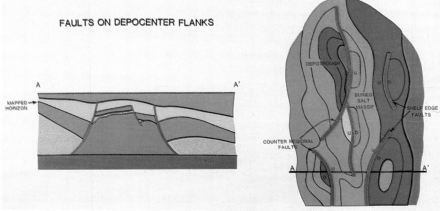

Figure 5.

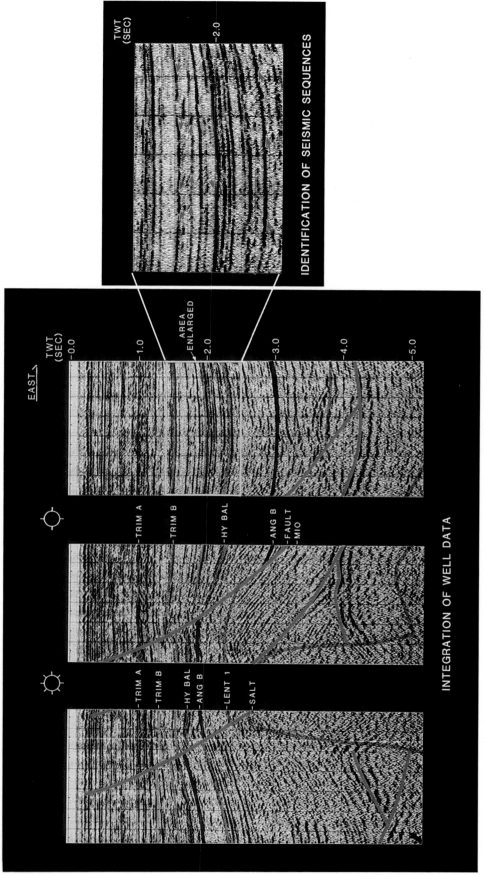

Figure 6.

interpret these faults, the explorationist must understand the genesis of different types of salt-related faults and the effect that they have on the surrounding sediments. The majority of salt-related faults are listric normal faults that sole out into the salt or overpressured shale. The amount of throw on the faults, and the direction of strike and dip are related to several factors, including: the relationship to different types of salt bodies, the stage of salt movement at which faulting was activated, and basin geometry.

The major types of faults related to salt movement are (Figures 2-5):

- Salt roller-related faults
- Tensional faults above salt features (crestal faults)
- Radial faults associated with diapirs
- Faults on depocenter flanks related to salt withdrawal
- Shelf edge growth faults

The salt roller-related fault develops during the first stage of salt movement. As salt moves laterally and separates into walls, faults commonly develop on the basinward side of the salt wall. The fault soles out into the next downdip salt wall and this brings about the collapse of the graben between the two salt walls. As the salt continues to localize, it will begin to climb the upthrown side of the fault, and the fault block rotates. This type of fault is very common where there is a basinward or seaward dip on the base of the salt, such as offshore Gabon or in the Campos basin offshore Brazil.

Crestal faults (tensional faults above salt features) develop during the second stage of salt movement, when salt starts to rise in the sediment section. The rising salt produces extension in the overlying sediments, resulting in collapse grabens. Collapse grabens remain active during the prediapiric stage, and sometimes serve as avenues for vertically flowing salt to form stocks and diapirs.

Radial faults form in response to diapiric growth. Radial faults develop because the intruding salt pushes the sediment apart. These tensional faults radiate from the intrusion in the same way that hairline cracks radiate from a bullet hole which passed through a glass plate.

As the salt flows into separate diapirs and stocks, it evacuates other areas. As the evacuation occurs, faults develop when the overlying sediment relates to infill the void left by the retreating salt. These faults develop along the sides of the salt walls, as the wall localizes into separate domes and diapirs. These faults form the edges of major depocenters, and are the normal up-to-the-coast growth faults that are present on passive margins. They may also form down-to-the-coast growth faults that serve as barriers to seaward sedimentation.

These down-to-the-coast faults are also called counter regional faults, which are prevalent, for example, in both onshore and offshore Louisiana.

The last major fault type to be discussed is not directly related to salt movement, but does form due to the presence of salt features. If a salt wall or elongate feature remains a positive element, then the seaward side of the wall commonly acts as the shelf-slope boundary. It is common in clastic dominated basins for growth faults to develop on the shelf break, due to deltaic progradation overloading the shelf edge. Therefore, even though there is not a direct link between the salt and fault, a cause and effect relationship nevertheless may exist.

Interpretation procedure. In salt-controlled basins, it is important for explorationists to reconstruct the geologic history of the area, including the effect that salt has had on paleogeomorphology and sedimentation. The explorationists' job, therefore, is threefold: first, all available exploration information must be integrated to determine key seismic horizons to interpret and map; next, the structural and stratigraphic framework of the area must be reconstructed by making a series of key maps and cross-sections; finally, the effect salt tectonics and sedimentations have had on hydrocarbon potential must be evaluated.

To make a complete and geologically sound interpretation of seismic data, it is necessary to integrate all available exploration data, most important of which are the well data. It is also important to utilize remote sensing data, gravity and magnetic data, published data, or other forms of available geologic data.

The interpretation process should begin with an identification of key seismic reflectors which then are interpreted and mapped. Ideally, these reflectors should be seismic sequence boundaries, which are unconformities bounding major depositional sequences. To determine the location of these sequence boundaries, several techniques can be utilized, chief of which is seismic sequence analysis.

Sequence boundaries determined by sequence analysis can be validated by integrating other exploration data (Figure 6). The example shown is from the Gulf of Mexico, where micropaleontology information obtained from the analysis of well cuttings is available. Because the major changes in these faunae are related to extinctions due to changes in sea level, the correlation of the micropaleontology breaks and unconformities picked on seismic is quite good.

Sequence boundaries can also be recognized on well logs. Again, the example (Figure 7) is from the Gulf of Mexico, which is a clastic dominated system. Sequence boundaries are indicated on the wells by major changes in depositional environment. An illustrative example would be the change from a shallow water deltaic sandstone at the top of sequence "A" to a transgressive marine shale directly above it. In this example, this sequence boundary results from an eustatic sea level change caused by glaciation during the Pleistocene.

By integrating the well data with the seismic data through the use of synthetic seismograms, seismic reflectors can be chosen that, when interpreted and mapped, can yield valuable information concerning the history of salt growth in an area. Once the key stratigraphic boundaries have been determined, the explorationist can start a study of the structural framework of the area.

From the previous discussion of salt growth and salt-related faults, it is clear that the structural framework of a salt-controlled basin is quite complex. It is important not only to recognize salt features and faults on seismic, but also determine the type of fault or salt feature and its effect on sedimentation. This requires going to a map view, and determining the areal extent of the structural features. Gravity data can be a useful tool at this time, especially if it is carefully integrated with the seismic and well data. Gravity data can be used to tell the location of salt features and help determine the true depth to the salt and the presence of caprock.

Because early salt movement can be controlled by presalt faults, the explorationist should also try to determine the presalt tectonic framework of a basin. These efforts will assist in determining the relationship between presalt faulting and the localization of major salt features. A study of the plate tectonics of an area and magnetic trends will aid in this presalt study.

A map showing the location of different types of salt structures might also aid in determining the presalt tectonic framework of the basin. Small, isolated diapirs, salt rollers, and nonpiercement salt features commonly indicate zones of thin salt deposition, whereas salt massifs and large diapirs indicate zones of thick salt deposition. Therefore, by separating the basin into dominant salt body types, the explorationist can make some assumptions as to the paleobathymetry of the presalt basin. This technique is especially useful in passive margins and rift basins.

Once the presalt and postsalt tectonic frameworks have been established, a series of structure maps on key seismic horizons should be constructed. A suite of structure maps will show the effect of salt movement on the overlying and surround-

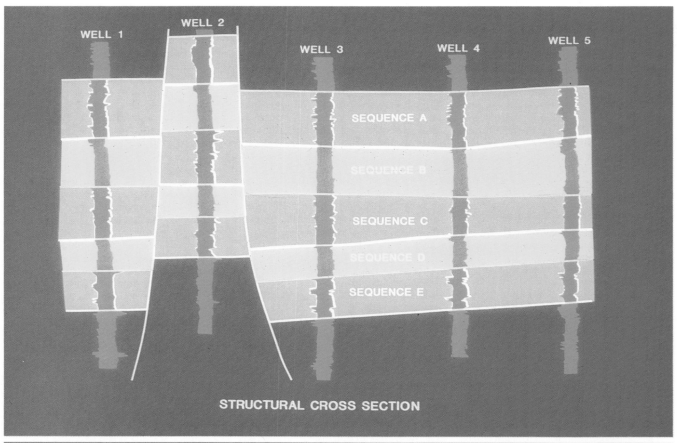
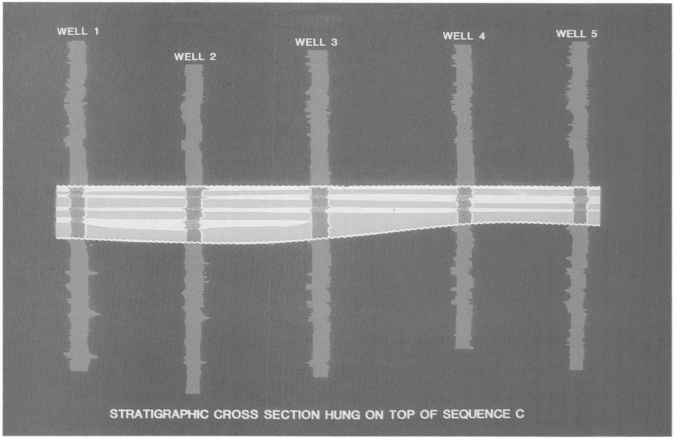

Figure 7.

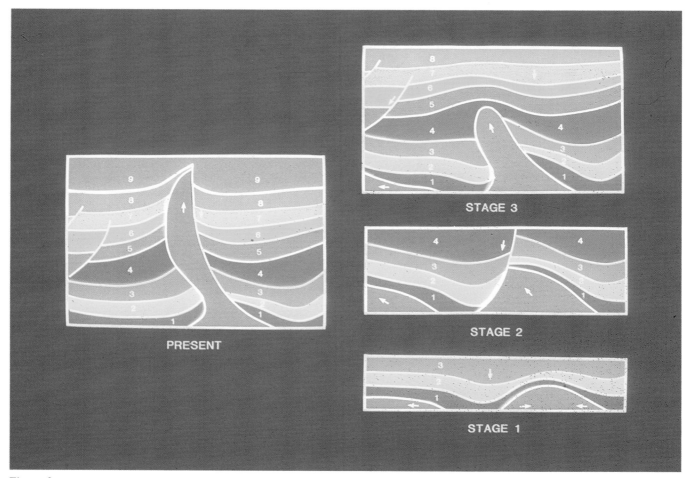

Figure 8.

ing sediment package. A suite of structure maps is necessary to identify the difference in structural control on deep versus shallow horizons. In complex areas, a reliable structure map may be difficult to construct due to a lack of a closely spaced seismic grid. If this is the case, a good understanding of the geologic models controlling salt growth will be needed to complete an accurate map.

After constructing structure maps, it will be necessary to develop an isopach map of each depositional sequence. Because salt commonly deforms through a predictable series of growth stages, isopachs should reflect these different patterns. The series of thicks and thins on the isopach map will highlight the locations of paleohighs and lows. Isopach thins will develop over the crest of the salt structures, and thicks in the old peripheral sinks, depopods, and depotroughs. Once this information is known, the explorationist can begin to determine sand fairways, probable areas for source bed accumulation, and other factors affecting the petroleum geology of the region.

In areas of complex salt growth, it generally is necessary to reconstruct the growth history along two-dimensional profiles. On seismic sections, this can be done by flattening horizons. Using this method, the explorationist attempts to reconstruct the paleogeography at the time of a certain depositional sequence. This method presumes that the top of the sequence was deposited horizontally or nearly so. This presumption is reasonable in most basins where the sedimentation rate is approximately equal to subsidence.

The example shown in Figure 8 illustrates how this method was used to understand the growth history of a salt diapir and its effect on sedimentation. Even though the piercement feature is presently related to a counter regional fault, one can see that the majority of deposition was controlled by a nonpiercement feature and normal, down-to-the-basin faulting. Production surrounding this diapir comes from sequence 4. Geological reconstruction reveals that at the time sequence 4 was deposited, cleaner sands should have developed on the downthrown side of the normal growth fault, on the southern side of the present day diapir. The drill has confirmed these findings, encountering thicker and cleaner reservoir sands on the southern side of the diapir.

With a suite of structure and isopach maps, well cross-sections, reconstructions, and key seismic sections, the explorationist can construct a series of paleodeposition diagrams for each prospective depositional sequence. A paleodeposition diagram integrates key stratigraphic fairways with important structural controls, yielding a diagram that can be used to identify possible locations for stratigraphic or combination traps.

A paleodeposition diagram therefore represents an integration of all previous work. Starting with an isopach map, the shape of the salt structures and the general trends of the stratigraphic fairways can be sketched. Then, by using the flattened reconstructions of the seismic data, a better description of the size and shape of each of the salt structures can be posted. Also from seismic data, information on syndepositional faulting and location of peripheral sinks can be added to the map. The well data and cross-sections are then analyzed to determine the known areal extent of the sand packages. Finally, after examining any remaining data available and using sound geologic sense, the other environments of sand deposition are added to the map.

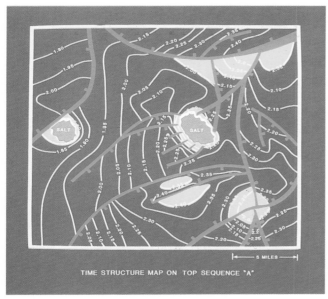

Figure 9(a).

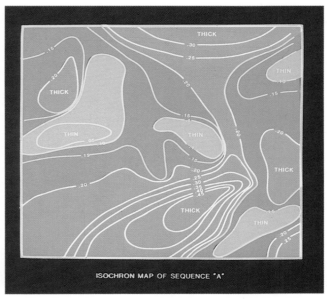

Figure 9(b).

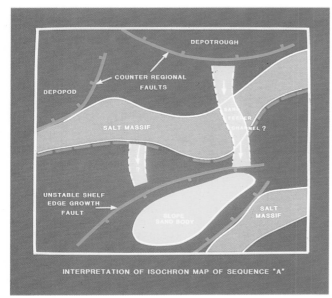

Figure 9(c).

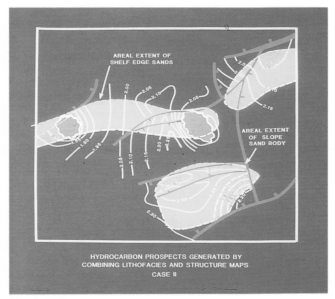

Figure 9(d).

The last step of the interpretation process is to combine all the maps and cross-sections to determine prospective trends. Figure 9 illustrates this procedure for one sequence of interest. The structure map on top of this sequence, combined with the isopach and paleodeposition diagram, can show potential sites of structural, stratigraphic, and combination traps. Key ideas and concepts can be highlighted to illustrate the location of subtle traps. By examining other sequences, decisions about source beds, migration pathways, and competent seals can also be made.

Ramifications of using this interpretation procedure. Because of the necessity of integrating all available data in finding subtle traps in a salt-controlled basin, it is difficult for a single explorationist to complete an interpretation of all data types. To overcome this problem, a multi-disciplinary team should be assembled to handle this type of interpretation. By combining several explorationists in different disciplines such as seismic interpretation, seismic processing, well log interpretation, and mapping, a powerful team can be utilized to arrive at better prospects.

Also, because of the geology in salt-controlled basins, the explorationist should always keep pace with the latest advances in data acquisition, data processing, and interpretation systems to help better image the subsurface. If exploration geophysicists can meet these challenges, they can more successfully hunt for the subtle trap, and become invaluable assets to their companies. **LE**

James Fox works for Phillips Petroleum Company in the Integrated Exploration Branch, Basin Analysis Section. He received his degree in geology with a geophysics option from the University of Missouri at Rolla in 1979. He has been employed by Phillips since, and has worked with salt controlled basins in many areas of the world, including the East Texas Salt Basin, Oriente Basin of Peru, the North Sea, offshore Brazil, offshore West Africa, and most recently as project leader of a regional study offshore Louisiana.

SEISMIC INTERPRETATION OF TRANSGRESSIVE AND PROGRADATIONAL SEQUENCES

By D. BRADFORD MACURDA, JR.
The Energists
Houston, Texas

The seas came in, the seas went out. This is the most enduring memory that many students have of a course in historical geology. The statement dulls our appreciation of the depositional systems resulting from transgressive and progradational events and leaves us with a static concept of the accumulation of the stratigraphic record. The result is that we are hindered in our exploration efforts aimed at finding the reservoir facies in transgressive and progradational systems, whether they are carbonate or siliciclastic.

To make maximum use of seismic data in exploring for the reservoir facies resulting from transgressive and progradational events, one should understand the processes inherent within them.

It is rare for sufficient energy to be present in most environments on a day to day basis to actually effect significant sediment transport and deposition of sediments. Therefore, one must look at longer time spans and understand the role of the hundred year floods, hurricanes, or cyclones in molding the materials we see. One must also look at the relative contrast of the rate of sedimentation versus the rate of accumulation; the latter is a fraction of the former. Research on the deposition of the shallow marine sediments off the Yellow River of China showed that the rate of sedimentation was tens of times higher than the rate of accumulation. Sediments were brought in each spring by the floods, resulting in the measurement of a high rate of sedimentation. Storms and currents redistribute the shallow marine sediments resulting in the rate of accumulation being only a fraction of the rate of sedimentation. An examination of shelf sediments globally found that most present day shelf areas (75 percent) are presently either sites of non-accumulation or erosion. The accumulation of the stratigraphic record is episodic and highly variable in time and space.

What governs the rate of accumulation? It is important to recognize the role of base level in sediment accumulation and the resulting variability of the stratigraphic record. Base level is an equipotential surface, up to which sediment can accumulate but not above. To many people, base level is sea level; this is not true. We find toplap and sediment by-passing associated with deep sea fans; something has resulted in the system prograding instead of both prograding and aggrading. Base level is a complex interplay of tectonics, sea level changes, current, and the rate of sediment supply. Alpine lakes can be sites of deposition; abyssal ocean floors can be sites of erosion. Thus base level fluctuates with respect to any point on the earth's surface. If it lies above the surface, we have the potential for sediment accumulation; if it lies beneath, we have erosion. Transgression across basin margins results from base level rising relative to the site: progradation occurs when we are no longer creating new space and sediments must be transported to deeper portions of the basin. If base level falls beneath the sediment-air and/or water interface, erosion and removal will occur.

When we explore for reservoir facies resulting from transgression or progradation, certain parts of these are of much greater interest than others; if we could identify them within a sequence (the basic unit of seismic stratigraphic analysis), we would be that much ahead.

The terms depositional systems (e.g., a barrier island and its relevant lithofacies) and depositional tracts (a combination of systems as the barrier island, the lagoon behind and the shallow offshore in front) can be used to describe sedimentary environments. Transgressive shelf deposits have been called transgressive system tracts (TST); shallow marine progradational events have been called highstand systems tracts (HST). We have generally failed to fully analyze the exploration potential of transgressive deposits. Using both outcrop and well information, studies from the Triassic of the Sverdrup Basin in the Canadian Arctic, have demonstrated the interrelationship between transgressive and progradational tracts. The transgressive units are relatively thin; they would thus usually be below the limits of seismic resolution. In the Triassic of the Sverdrup Basin, these were comprised of fragmental carbonates, sandstones, or phosphatic ironstones. As the transgression proceeded, the most active site of accumulation would episodically shift toward the basin margin and the rate of accumulation on earlier units would decrease to a very low rate, producing a condensed section. The condensed section would then be buried by the progradation of the highstand systems, producing a feature that has been called a downlap surface (DLS).

It is important to note that this is not a sequence boundary as it is a local, time transgressive surface within a sequence. Nine such transgressive-progradation sequences were recorded from the Triassic of the Sverdrup Basin. Field research, using the magnificent exposures of the Permian of southeast New Mexico, has shown the TST's and HST's on the outcrop; the slope of the basin margin was steep enough to compress the lateral scale so stratal relations stand out very clearly on a local scale. Seismic data shot across the same trend a few kilometers away where it is downfaulted showed the same configurations of Carlsbad, New Mexico. Thus, here one has the ability to compare the outcrop and seismic and see just what kind of lithofacies are present as well as comparing stratal and reflection configurations.

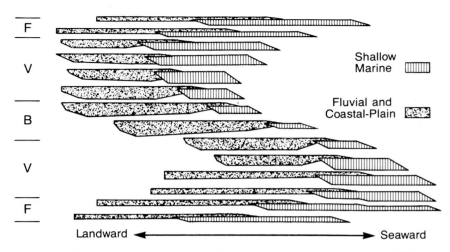

Figure 1. Episodic accumulations comprise the transgressive systems tract (TST) and highstand systems tract (HST). The units of the TST are backward stepping (B) toward the basin margin. The overlying HST shows vertical stacking (V) and forward stepping (F-progradation). (Courtesy T. Cross.)

These efforts point toward a model that we can test for exploration purposes. It is comprised of a transgressive systems tract which is a series of small scale progradational events, each subsequent one being further shifted toward the basin margin. This is illustrated in Figure 1. Here a series of depositional episodes (parasequences) are shown in an exploded fashion. Each consists of a fluvial-coastal plain and shallow marine facies. The most sand-prone facies for each of these would be at the junction of the two. The resulting geometry of these stacked units is a function of whether they are deposited over a gentle or steep gradient, the amount of shale or fine-grained carbonate, and whether coarser-grained sediments are present. These will determine the geometry and continuity of a potential reservoir. The condensed section can potentially provide a seal. The progradational system is comprised of a series of episodically formed units; subsequent units are formed in a more basinward position as shown in Figure 1.

It is our task as explorationists to evaluate both the TST and HST as to their internal configuration and the potential for siliciclastic sands and/or coarse grained carbonates being part of these units. We would like to stack as many potential sand prone facies above one another as possible. This would be where the transgression stops and the progradation begins as shown in Figure 1. If the formation of the TST and HST was driven by allocyclic events such as Pleistocene sea level changes, the occurrence and number can be expected and predicted in other areas. If they result from autocyclic events such as delta switching, they still have exploration interest but we are not able to make predictions beyond the immediate basin.

A seismic example of a TST-HST is shown in Figure 2. It comes from the northeastern Gulf of Mexico. Two sets of prograding reflectors are shown; each shows toplap and these represent sequence boundaries. Just above the lower sequence, we see two parallel reflectors and then downlap of the upper prograding reflectors onto these. The parallel reflectors represent the TST. There is a condensed section on top of these and the downlap is onto a DLS. The TST thins to the right (northeast); the turnaround point can be identified and thus the likelihood of the portion of the sequence with reservoir-prone facies enhanced.

Figure 3 shows a prograding lower Tertiary delta in the Porcupine Basin off Ireland. A flat lying parallel reflector marks the top of the chalk. The siliciclastics prograde from right to left (north to south). The mound above is part of a series of subsequent submarine fans. The prograding interval has low reflectivity and is often downplayed as to its significance. It marks a time of significant deepening of the basin. The progradation is episodic. The black and white wiggle trace section suggests this; it is more apparent in the color amplitude display in Figure 4. A monochromatic phase display of this same interval in Figure 5 brings out the episodic progradation quite starkly while a multicolored phase display at an expanded display in Figure 6 brings out the fine details. Use of this kind of display on prograding systems should help to determine the turnaround points between the TST and HST within a sequence and improve our exploration success.

These ideas are the results of the work of many people and not original herein. I am indebted to these people through their writings and many conversations.

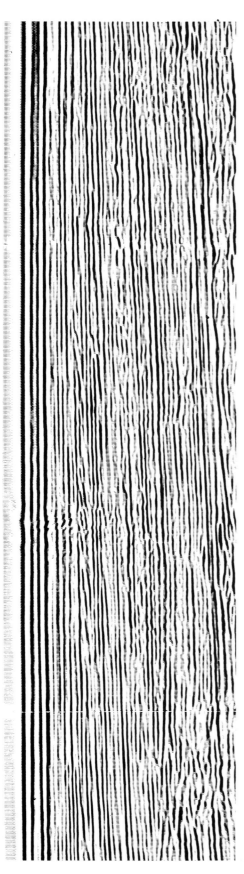

Figure 2. CDP record section from northeastern Gulf of Mexico showing two highstand systems tracts prograding to left with an intervening transgressive system tract. (Courtesy GECO, USA.)

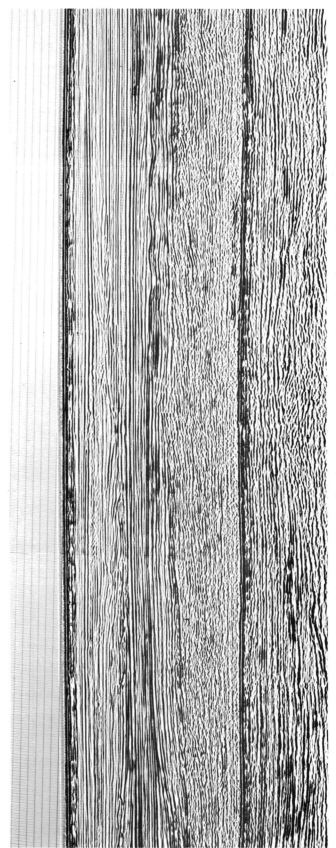

Figure 3. Wiggle trace display of lower Tertiary delta system prograding from north (right) to south. Porcupine Basin, offshore Ireland. (Courtesy Merlin Geophysical.)

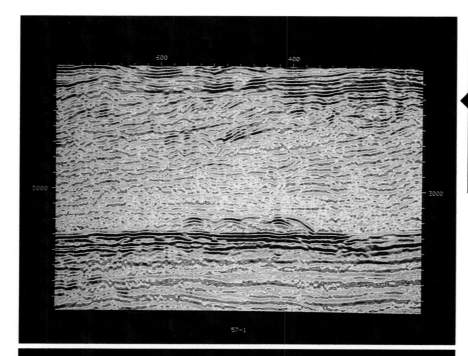

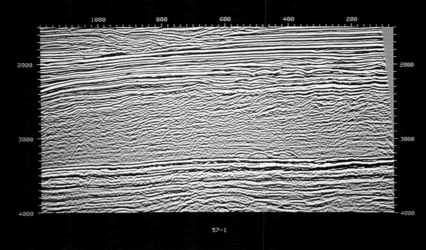

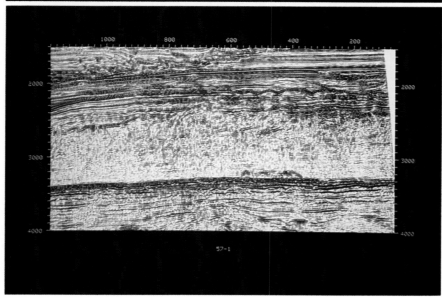

Figure 4. (Top) Color amplitude display of Figure 3. (Courtesy Merlin Geophysical and Landmark Graphics.) Figure 5. (Center) Monochromatic phase display of Figure 3. Figure 6. (Bottom) Enlarged multicolored phase display of Figure 3.

The ideas too must be tested and calibrated. We still have much to learn as to how to most effectively explore for reservoir facies in transgressive and progradational units.

The Porcupine Basin is a large, rift basin off the southwest coast of Ireland. Since its inception in the lower Cretaceous, the sedimentary fill has reached a maximum thickness of between 7-8 km. A wide variety of environments are present which offer many stratigraphic plays and traps.

The stratigraphic succession is divisible into 9 or 10 supersequences; there are 29 or 30 sequences in the Tertiary alone. Mapping all of these by conventional techniques based upon CDP record sections is a long, time consuming task. Once the sequence boundaries have been recognized and verified by loop tying, it is possible to use a workstation to greatly speed up the facies and mapping procedures. Displays of both amplitude and phase bring out significant differences in the depositional history and facies of the sequences.

A medial volcanic ridge marks the axis of the basin. The lower Cretaceous is marked by early basin fill and volcanic flows. A series of alluvial fans or fan deltas along the eastern basin margin offer a significant exploration objective. These are followed by some carbonate buildups in the upper Cretaceous which are also a significant exploration objective.

Deltaic progradation into the basin occurred in the lower Tertiary. A significant deepening of the basin occurred at this time. The development of a series of prominent submarine fan lobes provides a challenging problem to determine where the reservoir prone facies are best developed and where hydrocarbons would be trapped. Amplitude and phase sections reveal significant variations within the lobes and the subsequent sealing by onlap deposits. Mapping of the sequences within the fans demonstrates their structural attitude and how drainage would occur. This analysis was accomplished far more quickly with much greater insight using a workstation than conventional techniques would have afforded. **LE**

Seismic facies analysis: Pitfalls and applications in cratonic basins

By THOMAS L. DAVIS
Colorado School of Mines
Golden, Colorado

Various problems arise in relating seismic facies analysis concepts to geologic concepts and models developed from outcrop and well data. These problems are identified as:
• Origin of clinoform seismic patterns
• Number of different lithofacies models which give similar seismic response (seismic-facies models)
• Relation of reflecting horizons to time-stratigraphic surfaces
• Recognition of regional, inter-regional or basin-wide unconformities
• Application of seismic-facies analysis to cratonic basins

Clinoform seismic patterns. Figure 1 indicates that there are several depositional environments where clinoforms (inclined surfaces of deposition) occur. As a result, observation of clinoform seismic patterns may not be associated directly with a specific depositional environment and lithofacies without additional information to substantiate the interpretation. Clinoform seismic patterns are often associated directly with deltaic depositional environments. However, time-depositional surfaces in the delta front environment rarely exceed one-half degree of dip (a gradient of 50 ft/mile). It is unlikely that seismic events would be derived from time-depositional surfaces and that clinoform seismic patterns would be generated in a deltaic depositional environment or setting.

Peter Vail et al. (*Seismic stratigraphy and global changes of sea level, Part 5: Chronostratigraphic significance of seismic reflections,* AAPG Memoir 26, 1977) interpreted a deltaic environment and facies for the Woodbine formation, Tyler and Polk counties, Texas (Figure 2). Similar interpretations were published by G. R. Ramsayer (*Seismic stratigraphy, a fundamental exploration tool,* Offshore Technology Conference Proceedings, 1979) and O. R. Berg (*Seismic detection and evaluation of delta and turbidite sequences: Their application to exploration for the subtle trap,* AAPG Bulletin, 1982). In 1981, C. T. Siemers reported (*Deep-water clastic sediments — A core workshop,* Workshop No. 2 published by the Society of Economic Paleontologists and Mineralogists) that core studies indicate the depositional environment is not deltaic but slope related. Reservoir sandstones in the Woodbine in the area of the seismic line are turbidites and deep-water fans.

A calculation of dip of the clinoform seismic events from Figure 2 indicates 3 degrees (300 ft/mile gradient). This dip is indicative of a slope depositional setting. An interpretation of slope origin for the clinoform seismic patterns agrees with Siemers' core studies.

During lowstand conditions of sea level, deltaic systems could be superimposed on the shelf/slope break. However, Siemers pointed out from his core study that the upper surface of the Woodbine clastic wedge is not a lowstand unconformity. The Woodbine seismic sequence is not unconformity bounded as seismic sequence concepts would suggest.

Instead, the shelf prograded out over the slope until the sediment supply was cut off; then, deposition of the Austin chalk occurred.

Lithofacies vs. seismic facies models. A number of different lithofacies (controlled by depositional environments) can give similar response. For example Figure 3, from Alberta, shows a carbonate-dominated shelf margin whereas the Woodbine of Texas (Figure 2) is clastic. A geologic understanding of depositional environments and facies in clastic-dominated vs. carbonate-dominated environments is required for proper facies analysis. Expected potential reservoirs vary from fans at the toe of the slope in the clastic-dominated shelf margin to reefs at the top of the slope in the carbonate-dominated shelf margin. Seismic recognition of slope-related clinoforms must be accompanied by geologic modeling for proper facies analysis.

Reflections and time-stratigraphic surfaces. Reflections do not always come from time-stratigraphic surfaces. In shoreline deltaic environments, reflections may not come from them due to low angularity and the lack of acoustic impedance contrast. Reflections from shelf margin and slope-depositional environments appear to correspond to time-stratigraphic surfaces due to the nature and scale of sedimentation on the slope. Episodic disruption of pelagic sedimentation by density currents and downslope mass movements may cause changes in acoustic impedance and reflections which approximate time-stratigraphic surfaces.

Unconformity recognition. Unconformities are difficult to recognize on seismic sections, particularly in cratonic basins. Variability of acoustic impedance contrast along the unconformity surface, parallelism of events (disconformities) and the scale of erosion as it relates to seismic resolution all hamper seismic definition of regional unconformities.

The Gilby field in south-central Alberta is an unconformity trap that formed in response to a worldwide sea level drop in the Early Cretaceous, 112-114 mybp (Figure 5, page 21.) Production occurs from a valley-fill facies and from Jurassic and Mississippian age reservoirs trapped against the valley wall along the western edge of the valley. Maximum incisement or down-cutting related to the lowstand unconformity was 200 ft or 60 m.

Figure 6 is a seismic section across the Gilby Valley. The unconformity is interpreted from correlation with existing well control. Seismic events inside and outside the valley could easily be miscorrelated without subsurface geologic control. Events below the unconformity come from Jurassic and Mississippian strata whereas events within the valley are from

Cretaceous age strata. Amplitude is the main criteria for recognition of the valley fill. A higher amplitude is associated with the presence of the thick valley-fill facies.

Seismic facies analysis in cratonic basins. Seismic facies analysis procedures based on reflection configuration may not have applications to facies analysis in cratonic basin settings. The scale of stratigraphic sequences usually falls within the limits of seismic resolution. Amplitude and other waveform attributes are the main criteria for facies analysis (see *Seismic-stratigraphic facies models* by Davis, Geoscience Reprint Series 1, 1984).

As an example, one of the largest stratigraphic traps in the Rockies is the Bell Creek field in Montana (Figure 7). Recoverable reserves are estimated at 125 million barrels. Bell Creek produces from a shoreline/deltaic sandstone of the Cretaceous muddy formation that is generally less than 30 ft (8 m) thick. Geologic studies show that the Bell Creek sandstone was deposited on a paleostructural high prior to a 98 mybp sea-level drop. Erosion related to the sea-level drop resulted in a regional lowstand unconformity surface that caps the highstand regressive Bell Creek sandstone sequence.

Valleys preferentially developed in the paleostructural low areas that flanked the paleohigh. These valleys were filled during the subsequent sea-level rise. Seismic facies analysis of the highstand regressive shoreline/deltaic Bell Creek sandstone and the lowstand valley fill must occur by amplitude and paleostructural analysis. The valley-fill facies, which formed in the paleostructural low areas, is generally thicker and of higher acoustic impedance (amplitude) than the thin, porous Bell Creek sandstone.

Figure 8 shows a seismic line across the Bell Creek paleohigh. To the north of the seismic line, the Bell Creek sandstone undergoes a facies change to the valley fill and the lagoonal marginal marine facies of the highstand shoreline/deltaic sequence. This facies change created the trap at Bell Creek. The change from Bell Creek sandstone to valley-fill dominated facies appears as an amplitude change on the seismic section (Figure 8).

The Pembina field, Alberta, is a giant stratigraphic trap that formed by facies change within the Cretaceous Cardium formation (Figure 9). Structural and sea-level changes during Cretaceous time presumably led to hydrocarbon entrapment at Pembina. Seismic definition of the Cardium facies change occurs by subtle amplitude change (Figure 10). The amplitude of the Cardium sandstone event eventually dies out marking the eastern limit of the field through updip facies change from sandstone to shale.

In both the Pembina and Bell Creek

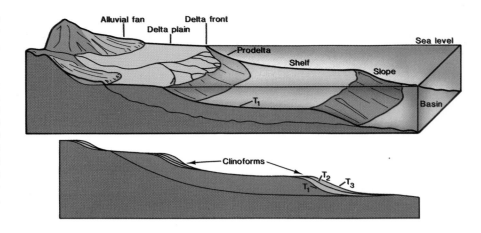

Figure 1. Depositional settings and clinoform development.

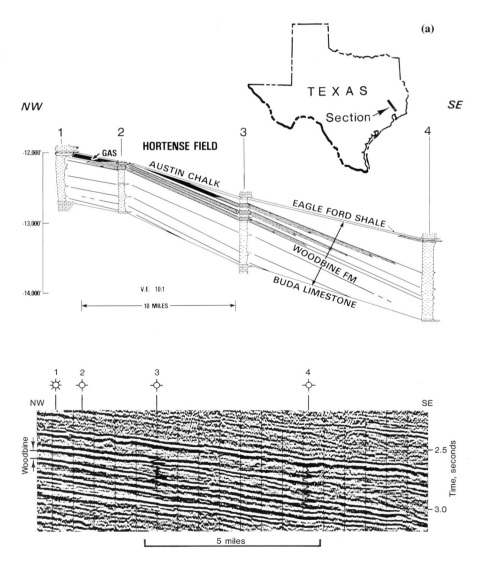

Figure 2. Cretaceous Woodbine stratigraphic section (a) and associated seismic facies (b) in East Texas. From Vail.

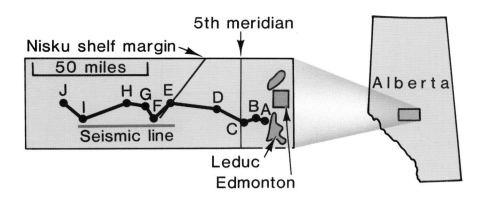

Figure 3a. Central Alberta.

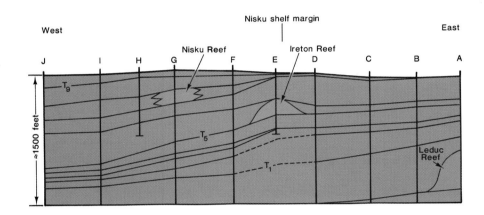

Figure 3b. Devonian Nisku/Ireton stratigraphic section.

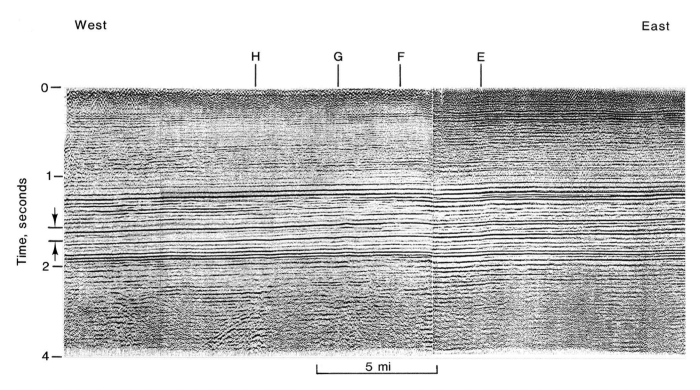

Figure 4. Seismic section from Central Alberta exhibiting shelf/slope sequence in Devonian Nisku/Ireton interval. Section courtesy of Chevron Standard.

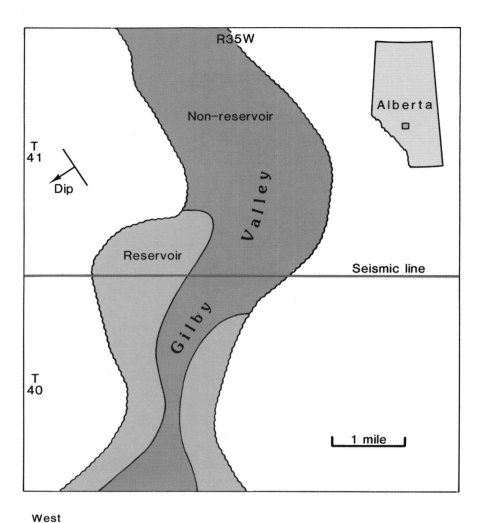

Figure 5. Gilby Valley (left) and related stratigraphic traps (below), Early Cretaceous, Central Alberta.

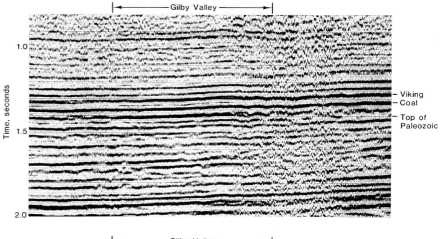

Figure 6a. Uninterpreted seismic section outlining the Gilby Valley. Sections courtesy of Chevron Standard.

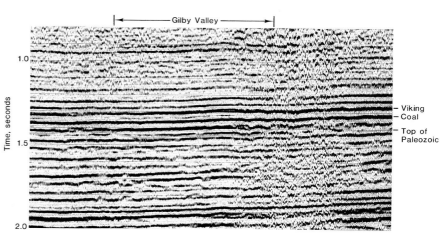

Figure 6b. Interpreted seismic section outlining the Gilby Valley.

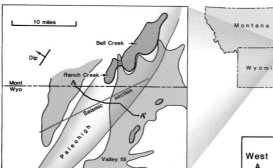

Figure 7a. Bell Creek field location. Courtesy of Bob Weimer.

Figure 7b. Stratigraphic cross-section.

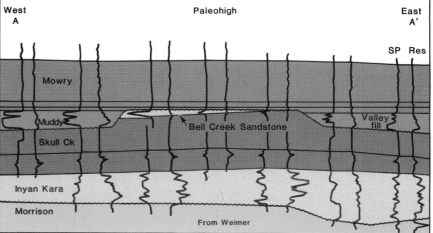

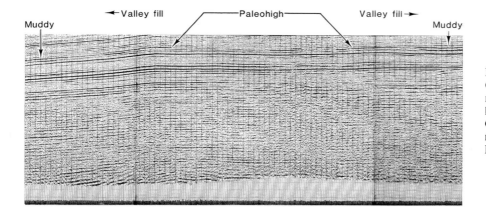

Figure 8. Seismic section across the Bell Creek paleohigh. Amplitude of the muddy formation is reduced across the high due to facies change to porous Bell Creek sandstone. Vertical scale is 600 ms/inch. Section courtesy of Petrel Exploration.

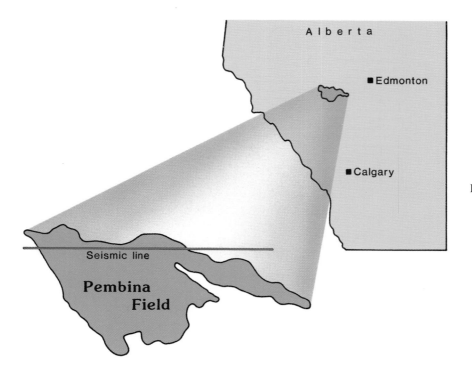

Figure 9a. Pembina field, Alberta.

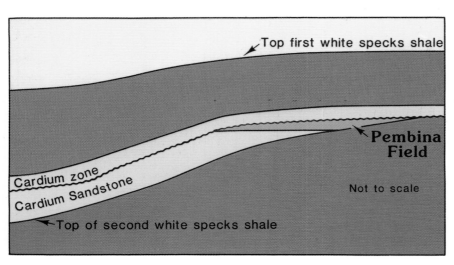

Figure 9b. Schematic cross-section.

(Continued on p. 37)

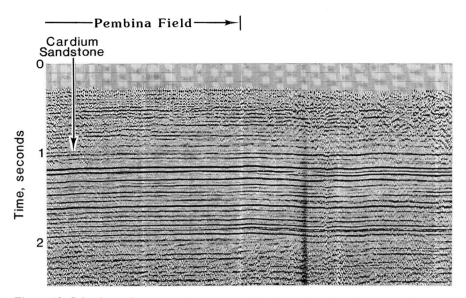

Figure 10. Seismic section across the eastern side of the Pembina field. Amplitude of the Cardium sandstone seismic event diminishes eastward and eventually disappears marking the eastern limit of the field. Courtesy of Mobil Oil, Canada.

fields, amplitude variations are the main criteria for recognizing facies changes associated with the reservoir and trap. Subtle structural control apparently influenced lithofacies distribution during time of sea-level change.

Seismic data must be integrated with geologic concepts and models to avoid pitfalls in seismic facies analysis. This integration can and will lead to new stratigraphic-trap exploration and development. LE

(Acknowledgments: Thanks are extended to Chevron Standard, Mobil Canada Ltd. and Petrel Exploration Consultants Inc. for allowing publication of the seismic data accompanying this article. Tenneco Oil Company is gratefully acknowledged for funding seismic-stratigraphic research at the Colorado School of Mines. The figures in this article were prepared through Tenneco's Research Contributions Program.)

Thomas L. Davis received a B.E. in 1969 from the University of Saskatchewan, his M.Sc. in 1971 from the University of Calgary, and a Ph.D. in geophysics from Colorado School of Mines in 1974. From 1971 to 1973 he served as a geophysicist in Calgary with Amoco Production Co., was an assistant professor of geophysics at CSM from 1974-77, and assistant professor of geophysics at the University of Calgary from 1977-79. In 1980 he rejoined CSM and is currently professor and assistant department head. Davis consults and lectures worldwide in the field of seismic interpretation.

Seismic interpretation of reefs

By ARMIN K. KUHME
Canterra Energy
Calgary, Canada

In the early years of exploration — the 1920s (when oil discoveries were made in the Mississippian formations of Turner Valley) — the Devonian formations of the western Canadian sedimentary basin presented a dilemma. They were not accessible beneath the Turner Valley discoveries because of depth and, where they were shallow, they were not accessible because of geographical remoteness. Technological advances in drilling equipment, settlement of remote northern areas, and an increased hunger for petroleum energy resolved the Devonian dilemma in the 1940s.

Although production from the Devonian dates back to 1944 when oil was discovered in the Princess field in southern Alberta, the treasure that lay far beneath the Paleozoic unconformity was a sleeping giant until 1947 when substantial reserves were discovered at Leduc by Imperial Oil Exploration. Since then, the Devonian formations have become the most prolific hydrocarbon producers, yielding 60 percent of Canada's oil and 25 percent of its gas.

In the area under discussion in this paper (where the Alberta, British Columbia, and Northwest Territories boundaries converge in Figure 1), the Middle Devonian Keg River is the most prolific petroliferous formation. Field work done by Shell Oil geologists in the late 1940s (on the heels of the Alaska highway construction) gave recognition to the Middle Devonian carbonate shale-out and hence reef development. In the gas-prone area to the west, the first wells were drilled, all in 1955, into the Middle Devonian by Gulf States Oil, Imperial Oil, Socony Vacuum and Shell Oil. In 1958, Gulf States made the first discovery and in 1964 Pacific Petroleums wells went on production. The major players in the '60s and '70s were Pacific Petroleums, Chevron, Texaco, BP, General American, Shell, and Quintana. Rising petroleum prices caused renewed interest by Mobil Oil (formerly Socony Vacuum) in the area and it became a major player in the late '70s/early '80s.

In the oil-prone area to the east, pinnacle reefs were first recognized in the Rainbow evaporite basin by Banff Oil in 1965. In the Shekilie evaporite basin, the first well was drilled by United Canso Oils in 1961. The first discovery was made by Del Norte Oils in 1969. Hudson's Bay Oil and Gas became the dominant player in the '70s. Mobil, Petro-Canada, Strand Oil & Gas, and Canterra Energy became significant players in the '80s.

Geology and reserves. The hydrocarbon-bearing Middle Devonian Keg River, which is also known (depending on locality) as the Pine Point or Presqu'ile, is the formation under investigation in this paper. The lower Keg River forms a carbonate platform on which upper Keg River reefs are developed. These reef-building carbonates are deposited over a large area of the western Canadian sedimentary basin — from the United States into the Northwest Territories — and from outcrops in the Rocky Mountains to subcrops beneath the prairies.

Figure 1 shows the geographical relationship of the carbonate shelf; the oil-prone restricted evaporite basins nestled in the carbonate shelf to the east; and the shelf edge and pinnacle reefs along the margin of the gas-prone open marine shale basin to the west. Dolomitized hydrocarbon-bearing Keg River reefs are developed in these environments. There are two types of reef associated with the open marine Clarke Lake shale basin.

First, the carbonate shelf edge is dolomitized and reefal. The Keg River forming the shelf edge reef is overlain by the Middle Devonian Slave Point formation which can also be dolomitized and reefal. When their shale-out edges are coincident and dolomitized (as in the geologic cross-section in Figure 2), the distinction between the two formations becomes difficult. In some cases, a thin band of Watt Mountain shale has been used as the demarcation between the two; in other cases, the overlying limestone has been named Slave Point and the underlying dolomite has been named Keg River. Both formations become thin far into the basin, and there they are together referred to as simply Middle Devonian Carbonate. Figure 2 illustrates the relationship between dolomitized shelf edge reef and pinnacle reefs in the shale basin along the north-south edge; whereas the geologic cross-section in Figure 11 illustrates the relationship between dolomitized Keg River shelf edge reef, Keg River pinnacle reef, and limestone Slave Point shelf edge reef in the Sierra embayment.

An important formation to note in both cross-sections is the Klua shale which is a source rock, a seal, and an excellent seismic reflector. To the north of the area, this Klua shale becomes calcareous and is referred to as a Klua equivalent. The western promontory on this shelf edge reef near Clarke Lake was the site of the 1958 gas discovery by Gulf States. There are only two prolific producing situations on the shelf edge reef — at Clarke Lake and Helmet. Both are updip shelf promontories sealed by Klua shale. The remainder of the shelf edge reef tends to be water-prone because the geologic formations exhibit a prevailing regional west dip and the eastern updip seal consists of tight carbonates of the Keg River and Slave Point formations and this is not always an effective seal. Therefore most of the shelf edge reef, although large in aggregate

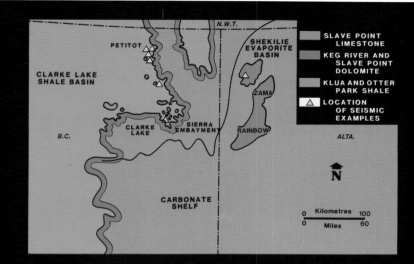

Figure 1. Map of the Middle Devonian system in the area of investigation.

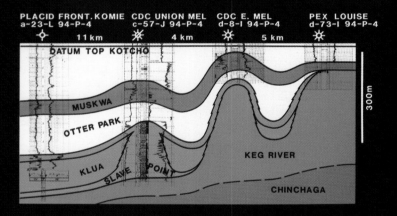

Figure 2. Stratigraphic geologic cross-section typifying the open marine shale basin environment and showing the relationship between shelf edge reef, pinnacle reef and shale basin in the Mel area.

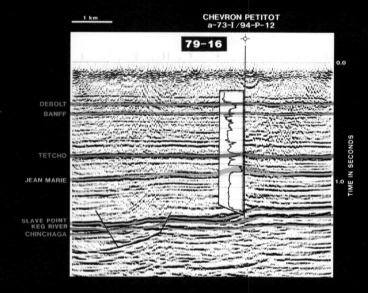

Figure 3. Migrated seismic line 79-16 showing regional relationship between north-south portion of the shelf edge reef (right) and shale basin (left).

areal extent, has reserves on the order of only 10 billion cu ft per section per well. On the shelf edge reef, 54 gas wells have been drilled to date to an average reef depth of 2,000 m. Net Keg River pays average 20 m; porosity is eight percent; water saturation is 20 percent; areal extent varies between 12,000 hectares (for the Clarke Lake field in the west) to 250 hectares in the north; the average Keg River reef build-up is 300 m (100 ms) and coincident average Slave Point cap is 100 m (40 ms); total proven recoverable initial reserves are 2,100 billion cu ft; total produced reserves are 1,600 billion cu ft; average flow rate is one million cu ft/day per well; the water-gas ratio is 50 bbl per million cu ft.

Second, dolomitized Keg River pinnacle reefs are developed basinward from the shelf edge reef in the Clarke Lake shale basin open marine environment as indicated in Figure 1. Most of these pinnacle reefs are developed in the shelter of the Sierra embayment. Some pinnacle reefs are developed north and west of this embayment; however, in this exposed environment, dolomitization is limited to within approximately 8 km of the shelf edge reef. Beyond this limit, Keg River pinnacles do exist but are not dolomitized. Figure 2 illustrates the relationship between limestone pinnacle reef and shale basin. Figure 11 illustrates the relationship between shelf edge reef, dolomitized pinnacle reef and shale basin within the Sierra embayment. All pinnacle reefs are encased in shales of the Klua and Otter Park formations which provide an excellent seal; many pinnacle reefs have a limestone (Slave Point) cap. On the pinnacle reefs, 31 gas wells have been drilled to date to an average reef depth of 2,100 m. Net Keg River pays average 85 m; porosity is 10 percent; water saturation is 12 percent; areal extent varies between 14,000 hectares (for the Yoyo field which is anomalously large) to 100 hectares (for the Gote field in the north) but averages around 250 hectares; the average Keg River reef build-up for dolomitized pinnacles close to the shelf edge is 300 m (100 ms) and the average reef buildup for limestone pinnacles far from the shelf edge is 200 m (70 ms); total proven recoverable initial reserves are 2,600 billion cu ft; total produced reserves are 1,500 billion cu ft; average flow rate is nine million cu ft/day per well; the water-gas ratio is 3 bbl per million cu ft. Therefore it can be said that pinnacle reefs in the open marine environment are prolific gas producers and much more rewarding but more difficult to find than shelf edge reef.

In the restricted evaporite basins, oil-prone dolomitized Keg River pinnacle reefs are encased in anhydrites and salts rather than shales. There the lower Keg River forms a carbonate platform on which upper Keg River dolomitized pinnacle reefs are developed. In addition, the Zama formation thickens and becomes dolomitized, porous and oil-bearing over Keg River reefs. On the pinnacle reefs, 80 oil wells have been drilled to date to an average reef depth of 1,800 m. Net Keg River pays average 35 m; porosity is eight percent; water saturation is 15 percent; areal extent is around 60 hectares; average Keg River reef build-up is 120 m (50 ms); total proven recoverable initial reserves are 20,000,000 bbl; total produced reserves are 12,000,000 bbl; average flow rate is 120 b/d per well; the water-oil ratio varies between .1-2.

Modeling and interpretation. If reefs are the treasure chest in this area, then seismic is the key. Rising oil prices in the late 1970s and early '80s caused renewed exploration interest in this area. Canterra shot extensive seismic surveys and drilled numerous wells at that time. Seismic was shot during the winter because the frozen muskeg allows access at that time. Usual acquisition parameters were:

- Dynamite source
- Two-hole shot pattern to attenuate troublesome ground roll
- Forty-eight or 96 trace bilateral spreads, 1,600 m to a side
- Nine in-line geophones over 34 m
- Twelve hundred percent coverage

Processing parameters included vigorous statics corrections to compensate for irregular muskeg overburden. Migration of the data was not seriously considered since tectonic disturbances are limited to a few normal faults with small throws.

The shelf edge reef, because of its water-prone reservoir, is considered to be only marginally economic and was not an exploration target. However, two seismic examples are included to show the regional relationship between the shelf edge reef and the shale basin. Seismic in Figure 3 is located in the north of the area and was shot perpendicular to the shelf edge reef where a well (Chevron Petitot a-73-I/94-P-12) is located. The large acoustic impedance of the reef carbonates overlain by a thick shale unit produces a clean, well defined reflection which drops off sharply (100 ms) from shelf edge to basin. Seismic in Figure 10 is located in the Sierra embayment and shot perpendicular to the shelf edge reef. The overhanging Slave Point carbonate formation produces a clean, well defined reflection which ends abruptly at the basin edge. An interesting feature to note is the Klua reflection which continues underneath the Slave Point overhang.

Pinnacle reefs, because of their prolific hydrocarbon content, were the real explo-

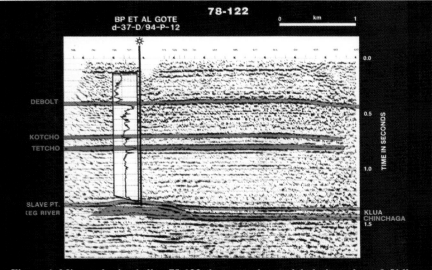

Figure 4. Migrated seismic line 78-122 shot over the modeled pinnacle reef. Yellow color indicates porous gas-bearing dolomite deduced by the modeling procedure.

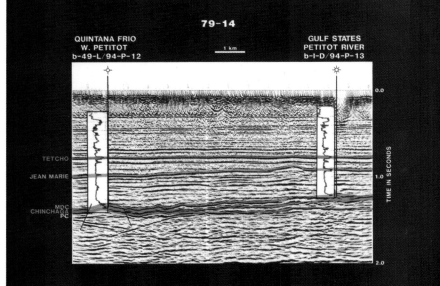

Figure 5. Migrated seismic line 79-14 showing a horst block (left) in the open marine shale basin environment.

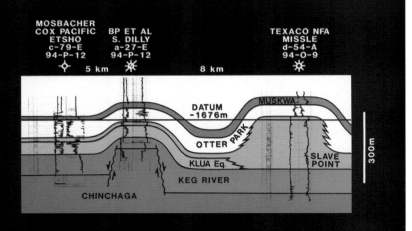

Figure 6. Structural geologic cross-section showing the relationship between a horst block and a limestone pinnacle reef in the Petitot area.

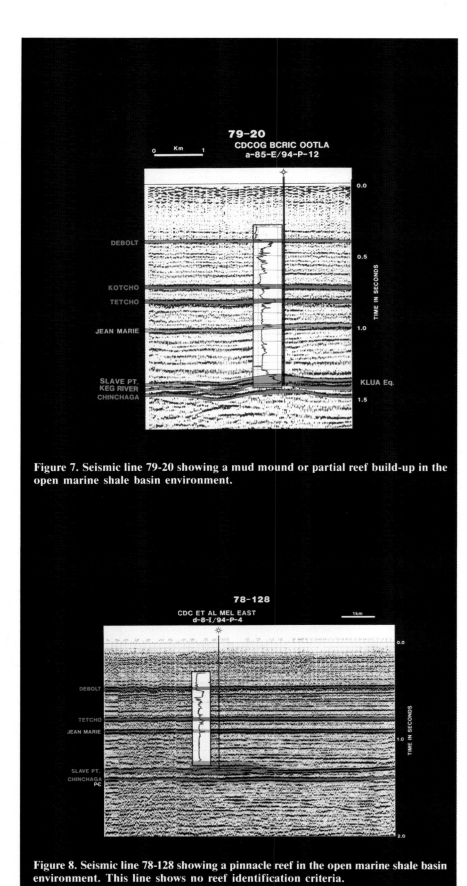

Figure 7. Seismic line 79-20 showing a mud mound or partial reef build-up in the open marine shale basin environment.

Figure 8. Seismic line 78-128 showing a pinnacle reef in the open marine shale basin environment. This line shows no reef identification criteria.

ration target. Seismic models were generated from geologic cross-sections over existing, economically producing reef anomalies. Seismic responses from these models were then examined and criteria for reef identification established.

All modeling was done with CGG's WEMOD package which is a forward modeling program based on vertical incidence, using the wave-equation method. A geological model — described in terms of depths and interval velocities (density option) that are derived from well logs — is input into the program. A time response is calculated corresponding to a fully migrated vertical incidence time section, and then demigrated using a wave-equation, finite difference algorithm. The result is equivalent to a zero-offset, stacked time section which can then be compared to an unmigrated CDP stack obtained from field acquisition.

Well BP et al Gote d-37-D/94-P-12 (proven recoverable initial reserves of 16 billion cu ft of gas) was used to construct a model for the open marine shale basin environment. From this model, the following criteria were derived for the identification of pinnacle reefs:

- Drape or isochron thinning from off-reef to on-reef between Tetcho and Slave Point formations.
- Shale discontinuity, i.e., where the reef is present, the Klua shale is absent and there is no reflection. To make the Klua shale clearly visible, polarities in modeling and interpretation were chosen such that a positive reflection coefficient corresponds to a seismic trough and a negative reflection coefficient corresponds to a seismic peak.
- Reduced amplitude and frequency at reef top which could be interpreted as porosity.
- Cycle onlap, i.e., the reef flank tends to give an oblique reflection.
- Attenuation of reflections below the reef.

Seismic in Figure 4 is shot over the same reef as was modeled. The criteria from the model can be identified on the seismic which was migrated to verify the reality of the cycle onlap criteria (which otherwise would be interpreted as a diffraction).

Pitfalls are present in any exploration effort because irreverent mother nature does not always comply with the orderly concepts present in the minds of men. Seismic in Figure 5 and geology in Figure 6 illustrate such a pitfall. Well Quintana Frio W. Petitot b-49-L/94-P-12 and BP et al S. Dilly a-27-E/94-P-12 are drilled on horst blocks. Basement horst blocks are present, especially in the northern part of the Clarke Lake shale basin. The difference between a horst block and a pinnacle reef lies in the thickness of the carbonate section; the former has a thin section of about 130 m, the latter a thick section of about 300 m. In addition, the Precambrian reflection of the horst block is continuous whereas in a reef the Precambrian, or lower, reflection is attenuated. Seismic in Figure 7 illustrates another pitfall. Well CDCOG BCRIC Ootla a-85-E/94-P-12 encountered an argillaceous limestone mud mound or an arrested reef build-up of only about 220 m that tested salt water from a thin band of porosity. The detection of porosity and hydrocarbon content still remains a challenge to the explorationist.

Seismic in Figures 8 and 9 illustrates the phenomena of directionality. The two lines were shot over the same anomaly (CDC et al Mel East d-8-I/94-P-4 with proven recoverable initial reserves of 24 billion cu ft) but in different directions. Line 78-128 which was shot perpendicular to the shelf edge exhibits no marked reef identification criteria, yet line 78-130 which was shot parallel to the shelf edge exhibits two distinct reef identification criteria (isochron thinning and cycle onlap). Seismic in Figure 10 and geology in Figure 11 illustrate a success story. Line 81-7 exhibits only one reef identification criterion, the discontinuity of the Klua shale reflection. Well Mobil Sahtaneh d-29-L/94-I-11 drilled on the anomaly is a pinnacle reef with proven recoverable initial reserves of 32 billion cu ft. This proves that one reef identification criterion alone is sufficient evidence for drilling . . . given that the explorationist's conviction in the soundness of his method is absolute.

The modeling procedure for the restricted evaporite basin environment at Shekilie (where pinnacle reefs are oil-bearing) generated the following criteria for the identification of pinnacle reefs:

• Drape or isochron thinning from off-reef to on-reef between Wabamun and Slave Point formations.
• Reduced amplitude and frequency at reef top which could be interpreted as porosity.
• Pull-down or thickening of the section between Slave Point and Precambrian formations.
• Specific areal extent, i.e., pinnacle reefs in the Shekilie basin tend to be about one-quarter section (60 hectares) or less in area. If the area appears larger on seismic, then the anomaly would in all likelihood represent a Fresnel zone response or a velocity variation.

Seismic in Figure 12 illustrates the validity of the reef identification criteria, all four of which can be seen on line 81-104 by the keen eye of the interpreter. Well CDC Shekilie 16-6-118-8W6 was drilled on the anomaly and encountered a Keg River pinnacle reef. The well has produced 72,000 bbl

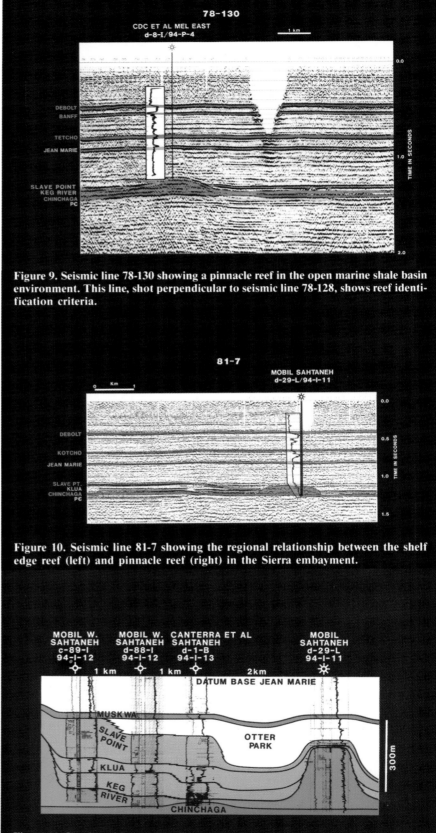

Figure 9. Seismic line 78-130 showing a pinnacle reef in the open marine shale basin environment. This line, shot perpendicular to seismic line 78-128, shows reef identification criteria.

Figure 10. Seismic line 81-7 showing the regional relationship between the shelf edge reef (left) and pinnacle reef (right) in the Sierra embayment.

Figure 11. Stratigraphic geologic cross-section in the Sierra area showing the relationship between the Keg River shelf edge reef (left), pinnacle reef (right), overlying Slave Point and the contemporaneous Klua formations.

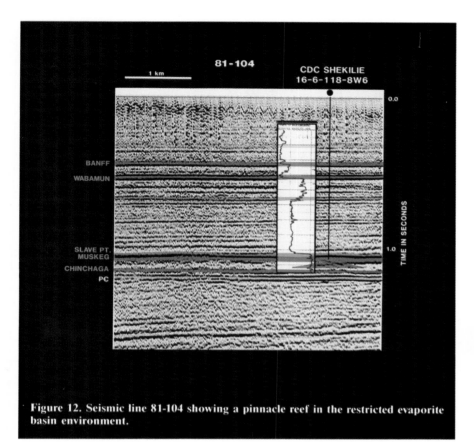

Figure 12. Seismic line 81-104 showing a pinnacle reef in the restricted evaporite basin environment.

to date at a flow rate of about 60 b/d. It is not exactly a boomer but proof positive that even smaller reservoirs can be found successfully with the proposed criteria.

A set of criteria for the identification of pinnacle reefs in an open marine shale basin environment and in a restricted evaporite basin environment has been established and used successfully in exploration. However, these criteria are not universally applicable and each basin will have a different set. The detection of hydrocarbon content and amount of porosity in reefs by use of the seismic tool is elusive at best and remains a challenge to the explorationist. **LE**

Armin K. Kuhme is senior staff geophysicist for Canterra Energy Ltd. of Calgary, Canada. In 1966, he received a B.Sc. Honors degree in geology and mathematics from Queen's University, Kingston, Ontario. Kuhme has gained his exploration experience in western Canada with Mobil Oil, Suncor and Canterra; in Germany with Prakla-Seismos; in Zimbabwe with the Zimbabwe Geological Survey; and in Indonesia with Suncor.

SEISMIC INTERPRETATION 5

The value of seismic amplitude

By ALISTAIR R. BROWN
Geophysical Service Inc.
Dallas, Texas

We have always observed amplitude in the seismic signal but, in the early days, we were concerned with its existence and not its magnitude because our objectives were only structural. Digital processing today generally seeks to preserve "true" amplitude so that stratigraphic inferences can be made from it and more subsurface information extracted from our seismic data. But how "true" are these amplitudes and how much can we infer from them? A workshop at the 1985 SEG Annual Meeting in Washington focused on this subject and concluded that our ability to control and understand seismic amplitude is far from perfect. This is undoubtedly true but there is nevertheless much valuable information in seismic amplitude that we must use to the full in our seismic interpretations and in our critical exploration and development decision making.

The most used interpretive application of seismic amplitude has been the *bright spot,* the amplitude anomaly interpreted as an indicator of hydrocarbon accumulation. But many interpreters today claim that the bright spot is unreliable and that it has fallen into disrepute. This view can be dispelled if we study all the pertinent character features of a suspected bright spot and use displays specially tailored to our needs. Much work on hydrocarbon indicators has been done in the Gulf of Mexico and four out of five of the illustrations in this article are from that area. Certainly gas reservoirs in Tertiary clastics no deeper than 2,500 m produce the amplitude anomalies most visible and amenable to study. However, advancing seismic technology is constantly permitting extension of bright spot studies into older rocks, at greater depths and in more difficult environments.

Figure 1 shows a Gulf of Mexico bright spot; in fact, it is two bright spots as detailed character studies indicate that there are two separate gas accumulations. These data have been migrated so that dipping reflections bear the correct relationship to each other, so that Fresnel zones have been minimized and amplitudes focused and so that diffractions from amplitude discontinuities have been collapsed. The data have also been wavelet processed to zero phase thus maximizing the wavelet energy into one symmetrical lobe. For high-amplitude reflections where the signal-to-noise ratio may be considered high like those in Figure 1, zero-phase character makes it significantly easier to relate individual reflection events to individual reflection interfaces in the subsurface. The data processing which generated what we see in Figure 1 also diligently preserved amplitude at all processing stages.

Now let us look closely at the character of the bright spot in Figure 1. Blue indicates positive amplitude, which, with the applicable polarity convention, means a decrease in acoustic impedance. Red indicates negative amplitude, which means an increase in acoustic impedance. Blue-over-red, with a high amplitude relative to surrounding reflections and thus presumably a high signal-to-noise ratio, indicates one low acoustic impedance bed, in this case a gas sand. Thus the four bright reflections indicate two gas sands. Let us look at other amplitude details to increase our confidence in this inference. The upper blue and red reflections dim at the same point updip and also at the same point downdip, which are different points for the lower reflection pair. The lower blue and red reflections are more widely separated, indicating a thicker lower sand. At the downdip limit of brightness is a *flat spot* indicating a fluid contact in the sand. It is unconformable, flat and bright. The fact that the flat spot fits between the top and bottom reservoir reflections and also occurs at their downdip limit of brightness is important evidence for validation of hydrocarbon fill. A fluid contact must separate hydrocarbon-saturated sand above from water-saturated sand below and thus must be a positive reflection coefficient. With the polarity convention here, this means a red reflection, which is how the flat spot appears in Figure 1. A flat spot will always be a trough if the data are zero phase.

Display is critically important to the seismic interpreter's recognition and study of hydrocarbon indicators. Color increases significantly the visual dynamic range and thus provides major benefits in studying detailed character features like those discussed above. The variable intensity blue and red coded to positive and negative amplitudes, respectively, as used in Figure 1, is a good, dependable color scheme but many others are useful. Experience is necessary to maximize the interpretive benefits of color. Figure 2 shows various displays of the hydrocarbon indicator of Figure 1. Four benefits of color for the study of amplitude detail are evident:

1. Color provides equal visual weight between peaks and troughs so that anomalies in either can be easily recognized and so that their amplitudes can be readily compared. Variable area/wiggle-trace display biases the interpreter's eye strongly to the peaks, making a comparison with the troughs very difficult.

2. High amplitudes, expressed as variable area peaks are often saturated when one trace excursion overlaps the peak on the adjacent trace. This does not occur with color if the display gain is properly chosen.

3. High amplitudes expressed as wiggle-trace troughs are not only difficult to see but have an excursion which places the amplitude maxima at a different subsurface location from where it belongs.

4. Dynamic range of amplitudes depends on horizontal scale for variable area/wiggle-trace display. Figure 2 demonstrates that six pixels per trace are the minimum necessary for a wiggle trace to display a useful amount of character. The dynamic range, and hence the character, of the color section is independent of horizontal scale.

Disadvantages of variable area/wiggle trace under items 2, 3 and 4 above all result from using the horizontal dimen-

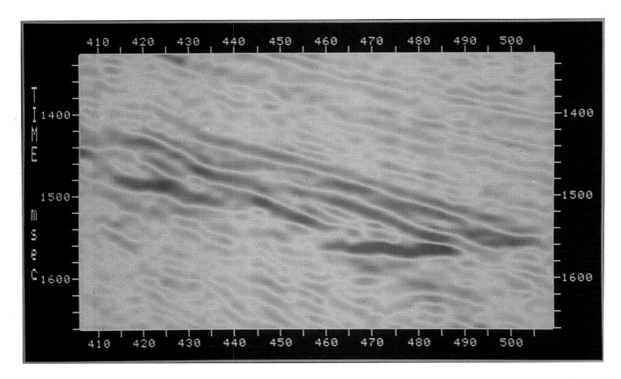

Figure 1. Gulf of Mexico bright spot showing the zero-phase response of two gas sands characterized by a blue-over-red reflection pair.

sion of the seismic section for both spatial position and amplitude of the signal. By using color these are segregated.

Figure 3 shows data from offshore California in which several separate hydrocarbon accumulations are discernible. Here the color scheme has been designed to increase visual dynamic range even further; gradational cyan has been added for the highest positive amplitudes and gradational yellow has been added for the highest negative amplitudes. (Cyan is the light bluish color; it is one of the three subtractive primary colors — yellow, cyan, magenta — contrasting the additive primary colors — red, green, blue.) Some of the hydrocarbon indications are labeled at the right. Each accumulation shows brightness and some show phase changes (polarity reversals). Several show prominent flat spots, unconformable with the structural reflections. Every indication demonstrates zero-phase character. Increasing gas sag is evident with depth and some reduced frequency can be seen. To emphasize the existence of the uppermost phase change, a structural horizon track from a reflection above the gas has been redrawn along the top reservoir reflection. This makes it clear that the reflection in the gas zone is a peak and outside it is a trough. The seismic section of Figure 3 shows so many good hydrocarbon indications that the interpreter must have a high confidence in their existence.

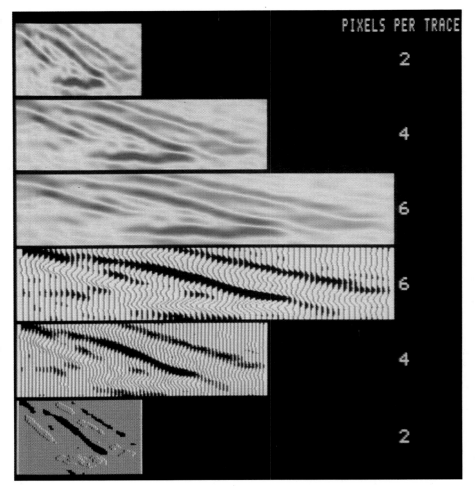

Figure 2. Visual dynamic range for studying bright spots is significantly increased by use of color and is not dependent on horizontal scale.

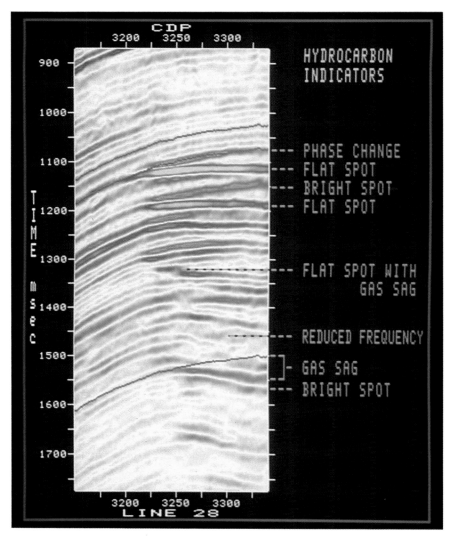

Figure 3. Hydrocarbon indicators offshore California are interpreted as the zero-phase response of several gas accumulations.

A further valuable utilization of seismic amplitude is on horizontal and horizon sections. A horizon slice is a section extracted from a 3-D data volume along one structurally interpreted horizon. It displays the spatial distribution of seismic amplitude over the chosen seismic horizon and is effectively the reconstitution of a depositional surface.

Figure 4 is a horizon slice through 3-D data from shallow water offshore Louisiana. The whole section is red because it is sliced along one seismic reflection which is a trough. The higher amplitudes, seen as darker reds, indicate two channels. Interpretation of this kind is a form of pattern recognition. The interpreter observes a pattern in the seismic amplitude and relates the shape in this map-style view to his geologic experience. The observable evidence of such a feature on a vertical seismic section is commonly rather subtle and would probably not generate any interest. When he sees characteristic shapes and forms in amplitude on a horizon slice, they are commonly dramatic so his attention is drawn to them. He must seek an interpretation.

As subsurface objectives become more detailed and geophysics plays a progressively more extensive role in petroleum development and production, interpreters would like to extract more quantitative information from seismic amplitude. Often our attempts become thwarted by fundamental ambiguities — what reservoir property is causing the amplitude variation? Porosity, hydrocarbon saturation and net pay thickness can all vary laterally over one reservoir and all affect amplitude in different ways. Other reservoir properties also affect amplitude but are more likely to do so for the reservoir as a whole.

Amplitude can contribute significantly to the location of development wells and to a reservoir description between existing well control, particularly in marine environments like the Gulf of Mexico. To a first approximation, a bright spot from a proven reservoir will be brighter where the reservoir contains more hydrocarbons regardless of which of the above properties is causing the brightening. Horizon slices are the best way to study the spatial variation of amplitude over reservoir interfaces. It follows that 3-D data employing the latest and best collection

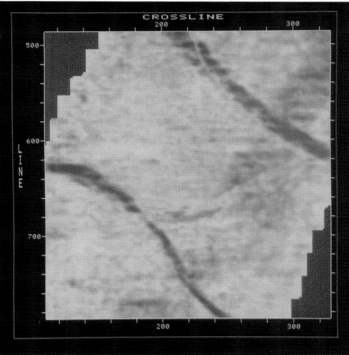

Figure 4. Horizon slice from 3-D data offshore Louisiana at a depth of around 2,500 m showing a pattern in amplitude interpreted as two channels.

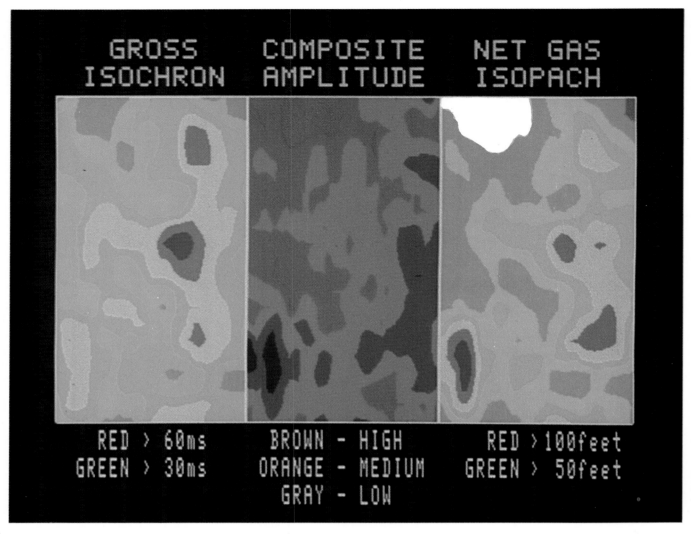

Figure 5. Mapping of net pay using a combination of seismic amplitudes and times over reservoir surfaces. Gross isochron (left) is the time difference between two horizon slices. Composite amplitude (center) after correction for tuning effects and calibration indicates net-to-gross ratio. Net gas isopach (right) is the product of the other two maps and a suitable interval velocity.

and processing technology are essential and this must include wavelet processing to zero phase. An interactive system and the flexibility of color display are highly desirable to manipulate the data and to provide a display of suitable dynamic range.

Given data and equipment of this kind, interpreters in the Gulf of Mexico are able to assess spatial variations within reservoirs under development. Figure 5 provides an example. Automatic horizon tracks provide time and crestal amplitude values over reservoir surfaces. Subtraction of times provides a map of gross reservoir thickness (left panel, Figure 5). Combination of gross time thickness with net-to-gross ratio and a suitable interval velocity gives net producible reservoir thickness (right panel, Figure 5). This approach is a pragmatic one incorporating several simplifying assumptions. However, it has been shown to work and to contribute significantly in developing optimum depletion plans.

For a thick reservoir, amplitude can indicate internal layering as well as lateral variations in those layers. Seismic inversion leans heavily on amplitude as it converts a section displaying the reflectivity of interfaces to one displaying the acoustic properties of subsurface intervals. Even qualitative seismic stratigraphy depends on amplitude as it distinguishes seismic sequences and attempts to define the seismic facies within them on the basis of gross character.

A look at the history of seismic interpretation indicates that amplitude has been used progressively more and more as technology has developed. Although yet far from perfect, seismic amplitude provides today's interpreter with displays pregnant with information. It is the responsibility of all interpreters to exploit that information to the full. 🕮

Alistair R. Brown is teaching and consulting in 3-D and other high technology seismic interpretation. He graduated from Oxford University in 1963 and then worked for the Australian Bureau of Mineral Resources. For the last 15 years he has been with Geophysical Service Inc. in England and Dallas. His principal interest is in experimental seismic interpretation and, particularly, the use of 3-D data, interactive methods and color for stratigraphic interpretation, field development and reservoir evaluation. He has presented many papers at conventions and also published extensively on seismic interpretation. He won the SEG Best Presentation Award in 1975 and was several times involved in the SEG/AAPG paper exchange. Brown's book Interpretation of three-dimensional seismic data *was published in December 1986. He is a member of SEG, AAPG, EAEG and is Chairman of the Editorial Board of* TLE.

SEISMIC INTERPRETATION 6

Discrimination between porous zones and shale intervals using seismic logs

By ROBERT C. MUMMERY
Teknica Resource Development
Calgary, Alberta, Canada

Reprinted from THE LEADING EDGE, January 1988
Society of Exploration Geophysicists

With recent advances in seismic processing and display capabilities, much emphasis has been placed on stratigraphic interpretation of seismic data. The shift from its structural counterpart began in the early '70s when Peter Vail of Exxon presented several classic papers on seismic stratigraphy. (An excellent summary of this work can be found in *AAPG Memoir 26*, 1977). The original emphasis of seismic stratigraphic interpretation was on reflection-character classifications and the identification of unconformities. This approach gained a large following during the mid-70s. It was during this period that many geologists began to work more closely with seismic data. New interpretation techniques have greatly increased our understanding of depositional environments and lateral facies changes within underexplored portions of sedimentary basins. Fortunately, the advances in seismic stratigraphy have reduced the number of unsuccessful wildcat drilling locations. However, dry holes are still being drilled.

While Vail's papers were being diligently studied and applied by most oil companies, Roy Lindseth published his paper (GEOPHYSICS, January 1979) on inversion of seismic data. Again, this paper represented a major change in seismic data processing. Since then, a variety of inversion techniques has been developed and applied to seismic data. Inversion has been very successful in providing detailed stratigraphic information in many of its applications. However, there have been some disappointments. This paper will focus on one of the common problems encountered by seismic interpreters. The problem relates to the differentiation between shale and porosity within a stratigraphic sequence. This problem can occur in a variety of stratigraphic and structural settings. Identification, remedial action and solution to this problem lie primarily with the interpreter.

The problem. Within a depositional sequence, compaction plays a major role in the increase of velocity with depth. The phenomenon has long been recognized and is well documented. Shale and mudstone sequences, under load, provide regular and predictable compaction profiles within individual basins. Variations between and within individual basins can be attributed to differences in rates of sedimentation, subsidence, and structural history. In general, the trends show a regular increase of velocity with depth. Reversals of this trend can be attributed either to structural reversals (uplift or reverse faulting) or to overpressured zones caused by permeability barriers to upward migration. The normal increase in velocity within shale sequences is attributed to the dewatering of the shales.

Sand or carbonate layers within these sequences will generally show trends similar to those observed in interbedded shales. However, in an interbedded sequence, shales will generally show a lower velocity compared to adjacent sand or carbonate layers. The higher velocities exhibited by sands and carbonates are primarily related to higher densities in these units compared with shales.

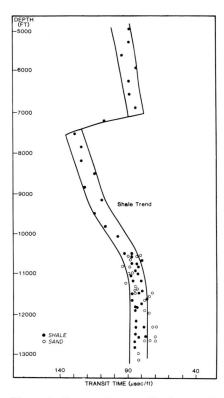

Figure 1. Compaction profile from well data in the northwest shelf of Australia.

Figure 1 demonstrates these relationships for an interbedded clastic sequence, utilizing data from a well located on the northwest shelf of Australia. This diagram shows the transit-time (inverse of velocity) values for the composite sequence plotted against depth as intersected by the well. The area outlined by black lines represents the velocity profile for shale units (solid dots) only. Open circles represent the values for the sand units. In this diagram, it is clear that shales show a narrow range of values at any given depth and define a good depth/velocity trend for this sequence. On the other hand, for the sand units, values exhibit more deviation for any specific depth. Sand values can be found on either side of the shale trend. To demonstrate this problem another way, Figure 2 shows a portion of the gamma-ray log for a well from the same region. Using gamma-ray data, it is not difficult to pick out the sand units. They are the major excursions to the left. However, since seismic data record only velocity-density variations (reflection coefficients), sand/shale differentiation in seismic data has the same ambiguity as a suite of sonic logs without companion gamma-ray curves.

In a clastic sequence, silts or thin sand units usually exhibit higher velocity than adjacent shales. The interpretation problem is caused by highly porous sand units which may have velocities similar to adjacent shales. In some cases, porous sands may even have velocities lower than shales. Since the goal of exploration and development is to identify and evaluate these sand reservoir units, the problem requires a solution.

The preceding remarks demonstrate that differentiation between a porous reservoir and a shale using sonic data alone, without the aid of a gamma-ray curve, or with a single inverted seismic trace, is extremely difficult. However, not all is lost. Another approach can be used to solve this problem.

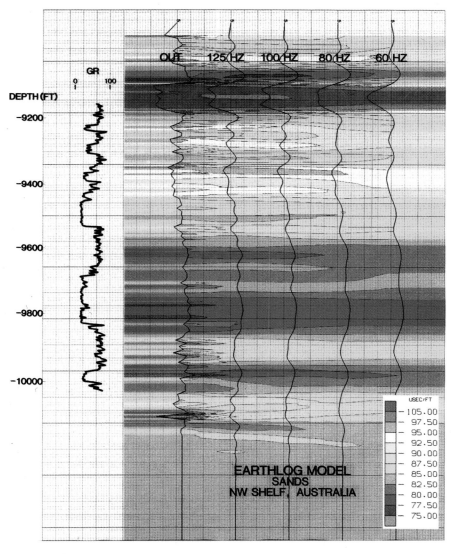

Figure 2. Sonic response in clastics.

Lateral extent. Within a stratigraphic sequence, individual units (sand, shale, or carbonate) will exhibit different areal dimensions based primarily on their individual depositional history. In most cases, the shale units are deposited under generally stable and global conditions. In many sedimentary basins, shales commonly make up at least 75 percent of the total stratigraphic column. In local areas, sands or carbonates may constitute a significant portion of the sequence. Exploration targets are generally located within or adjacent to transitional regions between predominantly shale and predominantly sand or carbonate. In sand or carbonate units, it is also common to find a wide range of reservoir quality (porosity and/or permeability). Figure 3 illustrates these relationships. Absolute values have not been shown on this diagram, since these can vary from area to area.

The critical clue is in the lateral extent of different units. On individual seismic lines, this technique can be utilized very successfully. Within a clastic or carbonate sequence, the lateral/areal extent of the different units (in decreasing order) can be predicted as: shales, sands/carbonates, and porous intervals.

Focusing on clastic sequences, some further explanation of these relationships can be made using Figure 3. Geological events can be classified according to their impact on the stratigraphic sequence. Global events represent uniform depositional environments which are stable for long periods of time (millions of years). For example, one centimeter of pelagic sediment can take 1,000 years to deposit. In a glacial environment a thin layer (millimeters) of carved clay represents one year of deposition. These examples generally represent deposition within a stable low energy environment. Coals, shales and mudstones are other sediments deposited under similar conditions.

Examples of higher energy depositions, which occur over millions of years, could be transgressive, or regressive clastic wedges. However, these wedges have two components: high-energy local units related to short period events, and lower energy regional units related to longer time periods. In Figure 3, the shales represent these latter units, and are commonly used for correlation purposes (shale markers). These stratigraphic units often exhibit uniform isopachs and facies. This results in a relatively uniform interval velocity for individual units.

On the other hand, the sands and silts between shale markers represent episodic local disruptions of the long-term global depositional environment. Fluvial, turbidite mounds, tidal sands and fluvial channels represent these types of high energy deposits. By contrast, these units have limited lateral extent and show rapid lateral isopach changes. They are often discontinuous and difficult to correlate using only borehole information. These units can exhibit rapid facies changes and thus interval velocities can show a wide range. Most clastic hydrocarbon reservoir units are found within these depositional environments.

These high energy events disrupt the lower energy regional sediments. But silts and poor reservoir-fine-grained sands can also be included with high energy deposits. Silts represent intermediate energy and can provide local correlations. These units have a more uniform composition and, locally can show constant interval velocities.

If these relationships can be observed in inverted seismic data within a prospect area, then potential reservoirs may be identified.

In processing seismic data prior to inversion, two critical factors must be addressed. First, it is critical to capture and retain relative changes in reflection amplitude, trace to trace. This is necessary to accurately map the lateral extent of individual units. Second, it is essential to understand the wavelet in order to make the best correlation with the sonic log data. In areas where well data is available, sonic information should be used for control. In a thinly interbedded sedimentary sequence, two factors can affect the thin-

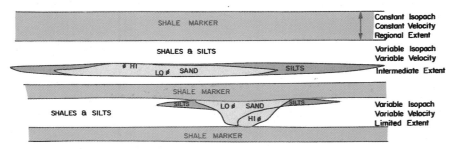

Figure 3. Diagrammatic relationship between porosity/velocity and areal extent.

bed signal interference and tuning effects tend to complicate the phase correlation problem. Furthermore, well data can sometimes be incorrect due to logging errors (hole conditions or tool malfunctions). Also, a few stratigraphic variations may have been intersected by the borehole which do not have sufficient areal extent to reflect the seismic signal (Fresnel zone).

The factors mentioned above warrant much more discussion than is possible in this paper. They have been mentioned as "red flags" which interpreters must look for when evaluating amplitude variations as part of the detailed stratigraphic analysis of seismic data. Assuming that these factors have been considered (and the data corrected for them as much as possible), let us return to the original problem.

A possible solution. Figure 4 shows the results of the inversion of some seismic data from Southeast Asia. Reflection coefficient information from seismic data has been recovered and inverted. This information has been quantified by the addition of low frequency (0-5 Hz) information. The results have been computer contoured and colored to enhance the information. Green colors represent the lowest velocities within the stratigraphic sequence; yellow and orange represent intermediate velocities; and blue and purple colors represent the highest velocities. Several units have been identified and correlated on the display. Low velocity zones, colored green, can be observed within the sequence shown in the figure. These units have two important characteristics.

First, there is little isopach variation

(Continued on p. 23)

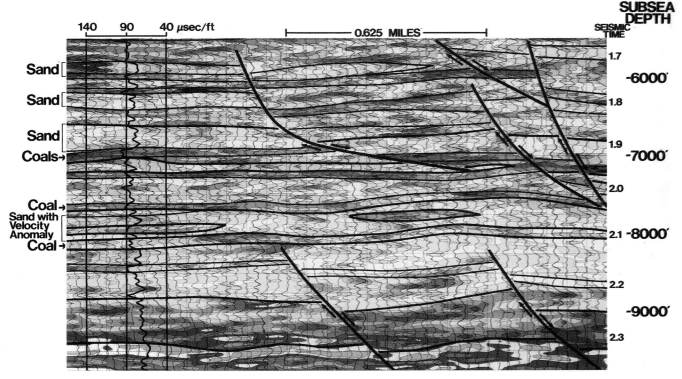

Figure 4. Color depth Seislog. Fluvial dominated depositional environment offshore Southeast Asia.

laterally across the display. Second, interval velocities have a narrow range. These units are identified as shales. On the other hand, the sand units generally show a higher velocity than the overlying or underlying shale units. This higher velocity is indicated by the pale yellow and orange colors. In contrast to shales, sands exhibit considerable variation both in isopach and interval-velocity value. Sand velocities range from well above (dark colors), to about equal to those of the shales (green). These variations are attributed to the higher-energy depositional environment of sand units. Several low-velocity units within the sand packages have been highlighted. These "lenses" have values similar to the shales. However, it can be readily seen that the lateral extent of these units is much more restricted. If one tried to evaluate the data from a single trace through one of these sand features, it would be impossible to distinguish the porous zone from shale beds above and below. However, these low-velocity zones change laterally into velocities more commonly associated with the sand sequence. Changes of this nature can often be very abrupt and are sometimes structurally dependent. For example, the gradation downdip from green to yellow may be related to changes in fluid content or porosity. Lower velocities (dark green) may be in part attributed to the presence of gas. The yellow colors downdip may be the result of secondary sedimentation within the reservoir's water-bearing portion.

In this example of inverted seismic data, many of the relationships shown in Figure 3 can be demonstrated. Coals and carbonaceous shales (green) represent regional and local stratigraphic units deposited under stable medium-term time periods. Sands and silts represent more local high energy units deposited during short period episodes. These sedimentary packages show a wide range of isopach values and interval velocities which reflect rapid facies changes occurring within these sequences.

I n summary, some knowledge of expected depositional environments is essential to correctly interpret lateral and vertical patterns exhibited on the inverted section. This information can come from borehole data, regional geological studies, or a previous regional seismic pattern recognition study. When incorporated with inversion data, detailed stratigraphic information can be extracted. The most important statement that can be made is that lateral and vertical changes are crucial to the interpretation. Overemphasis of single trace data near a well tie can be misleading. Similar methods are used when studying 48-trace records for differentiation between reflections, first breaks, and noise using pattern recognition rather than absolute amplitude or wavelet shape. LE

(Acknowledgments: I would like to thank Teknica for the use of its staff and equipment during the preparation of this paper; thanks are also extended to the anonymous donors of the data. Special thanks are given to Verne Street of Teknica for suggestions that were incorporated into the final version of the manuscript.)

Robert C. Mummery is vice-president, Interpretive Services, for Teknica Resource Development Ltd. He earned his bachelor's degree from the University of Western Ontario in 1968 and received a doctorate in geology from McMaster University in 1973. He spent the next nine years working for Amoco Canada, Home Oil, and American Hunter Exploration in a variety of development and exploration plays. Mummery joined Teknica in 1982 and currently is in charge of that company's interpretation projects with added responsibility for coordination between processing and interpretation teams. Other professional interests include basin evaluation studies; regional geologic and geophysical syntheses of onshore and offshore basins in North America, Indonesia, Pakistan, China, the North Sea, and Australia; field studies, seismic stratigraphy, and petrophysical studies.

Pragmatic migration:

A method for interpreting a grid of 2-D migrated seismic data

By COLIN O'BRIEN
London Geophysics
London, England

Areas of complex faulting cannot be interpreted satisfactorily using stacked sections, so in the absence of a full 3-D survey, it is necessary to interpret the migrated stacks. However, this leads to the well-known problem of data misties at intersections: migration changes the reflection time on dip lines but not on strike lines; therefore, if stacks tie, migrations won't. The traditional responses to this have been: ignore strike lines where misties occur, guess at the picks and contour through the misties, or tie the data around loops, introducing arbitrary faulting where "necessary." There is a fourth approach, described here, which results in accurately tied loops in virtually all circumstances. It has the added merit of leading to a finally contoured map which, on the average, is 90 percent correctly migrated and which can be extended to give a fully migrated map. This is the method which I term "pragmatic migration."

The basis of the method is that the returning wavefront is assumed to be represented accurately by the stacked seismic record and that intersecting stacks will tie accurately at all points. It presupposes that the processing has been accurately and consistently performed. The interpretation could be performed on the stacks, but details of fault planes and their associated stratigraphy cannot then be resolved. In addition, the usual assortment of diffractions, bow ties, sideswipe and so on cannot be completely disentangled. What is needed is a method of determining where the migrated sections tie, given knowledge of the intersecting positions of the stacks.

Figure 1 summarizes the relationships, at an intersection at shotpoint A, between the stacked section of a segment of a reflection, XY, and the migrated section $X_m Y_m$. If there were no dip, it would be possible to fold the stack at shotpoint A, lay it on the corresponding migration at that point and get an exact tie between the lines. However, the horizon time at the intersection T_b has changed to T_e, and the data from point B have meanwhile moved updip to point B_m, so a mistie can be expected. Now we know that the stacks of this line and one intersecting it at shotpoint A tie at time T_b, and we can find out where this time is now occurring by moving the folded stack in the updip direction along the migrated section, keeping the zero time lines of the two sections aligned until a match is obtained between them at point C. The intersecting line will now tie directly at the shotpoint A_m which corresponds to C, and it is no longer necessary to guess at the correlation between the migrated lines.

It should be clear that the data at point C have actually come from some unknown point D on the stack. If there is a rapid change of horizon character between B and D (say, some faulting or pinching out), then it will be impossible to get a precise tie with the crossing line. It will also be difficult to get an accurate correlation between the stack and migration, but it is possible to make a good guess by using horizons above and below the one under investigation. The assumption that the data at point D accurately represent the geology at point B is an integral part of the technique, and it is for this reason that the approach presented here is termed "pragmatic."

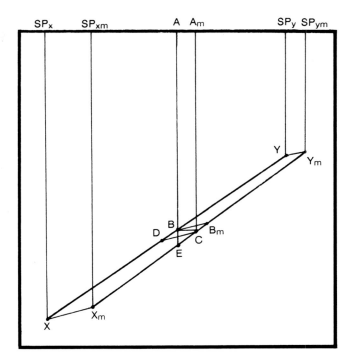

Figure 1. Relationship between reflections on corresponding stacked and migrated sections.

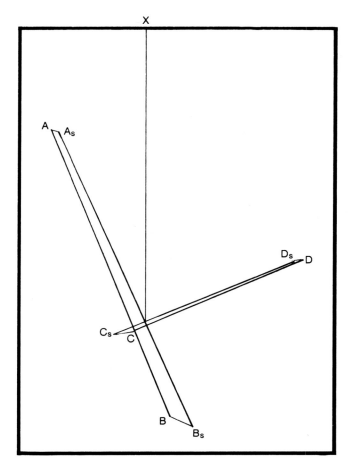

Figure 2. Inversion of true spatial relationships at a steeply dipping fault.

Point X is a particularly interesting case. If X represents the start of a reflecting segment, then any crossing lines between shotpoints SP_x and SP_{xm} will show a reflection corresponding to this horizon since it will exist on both intersecting stacks. However, the migrated version of this line will not have any reflection at all between times T_x and T_{xm}, and it will not be possible to find a tie. Conversely, the horizon exists only on the migration of the dip line between times T_y and T_{ym} if Y represents the end of a reflecting segment. Any crossing line which occurs between shotpoints SP_y and SP_{ym} will not show this horizon, so again it will not be possible to find a tie.

In extreme cases a complete inversion of spatial relationships can be obtained. Consider the situation in Figure 2, where AB represents the true (migrated) position of a steeply dipping fault and CD represents the true position of a bed dipping into it. On the corresponding stacked section the fault will appear at A_sB_s and the bed will appear at C_sD_s. A line crossing to the left of shotpoint X, where the reflections cross on the stack, will show the reflection from the bed actually beneath the reflection from the fault plane. Such data can be analyzed applying the pragmatic migration approach separately to the fault and horizon reflections, but should be treated with considerable caution.

The apparent difficulties which arise from the use of the method described here actually constitute one of its advantages; the interpreter is forced to acknowledge the problems which can occur when operating on migrated data. Furthermore, it is surprising how frequently the stack can provide an insight into the interpretation which cannot be gleaned from the migration, since the latter will not always be processed with the ideal parameters.

Method of application. The procedure requires a set of stacked sections and a set of migrated sections and is as follows:

• Establish the identity of the horizons to be picked and define the colors which will be used during the interpretation.

• Fold the stack at each intersection where movement of the reflections arising from the migration could be expected.

• For each intersection, lay the folded stack at the corresponding intersection point on the migrated section; then move the stack in an updip direction, keeping the zero time lines aligned, until it ties at the picked horizon. (Figure 3 shows horizon A aligned correctly, while horizon B requires being moved further to the right.)

• Mark on the migration the position where the two sections tie using a pencil of the same color as the horizon being picked.

• Proceed to the next deeper horizon, move the stack further in the updip direction and establish the migrated position as for the previous horizon.

• Continue this operation until the migrated positions of the intersections for all the relevant horizons are known.

The heading of the seismic record should look something like Figure 4 by the time this procedure has been completed. During the tying process it will be found that the shallow, flatter-lying horizons tie exactly, but the dipping, deeper horizons will show an increasing discrepancy with depth, depending on the dips and two-way times involved. Flat-lying horizons beneath dipping beds may also move. If the horizon has moved updip on both lines at an intersection, the tie must be made at the new position on *both* lines, even if by coincidence a tie can be obtained at the geographical location of the intersection. Ignoring the movement caused by migration will lead to incorrect maps at the con-

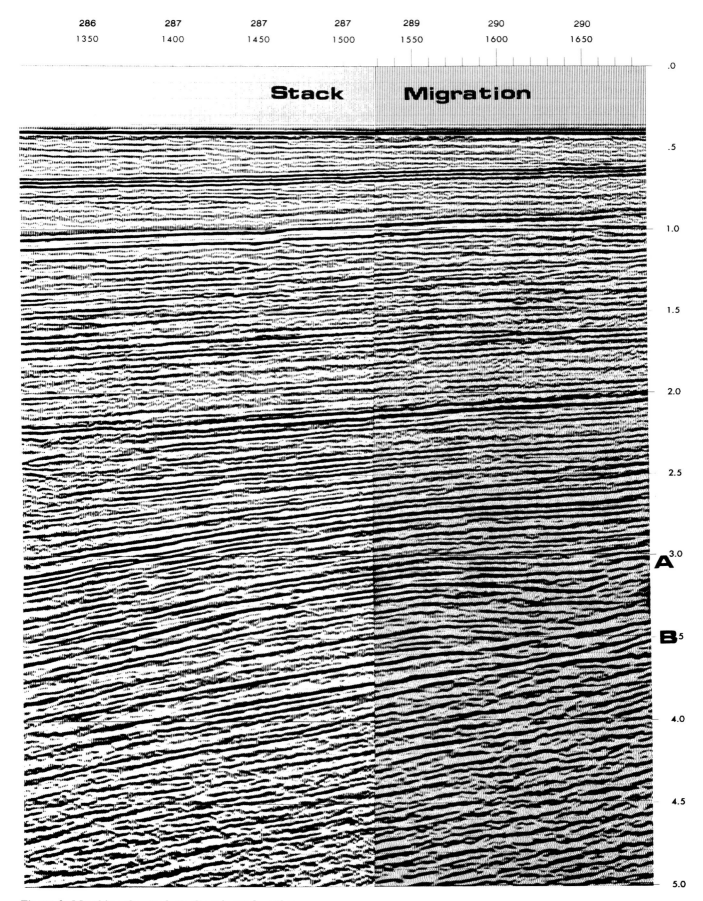

Figure 3. Matching the stack to the migrated sections.

touring stage.

The procedure described here is necessarily subjective, but it can usually be performed without trouble to the same degree of accuracy as a tie between two stacked sections. It will usually be found that the process is best performed with the downdip side of the intersection displayed on the stack and the updip side displayed on the migration, but it will occasionally be necessary to reverse this procedure.

As an alternative to preparing the complete set of migrated line intersection positions at the outset of interpretation, it is possible to work on an "as needed" basis during the interpretation, although this can sometimes prove to be inconvenient because of the need for access to the stacked sections. However, it may not be possible to treat complex areas in any other manner.

We now have a data set in which the migrated positions of each horizon are marked at each intersection, and the interpretation can be performed by folding the migrated sections at the newly determined positions of the intersections. In a grid where the lines follow true dip and strike, it is easier to fold only the strike lines since they will show no migration movement, and all horizons will have the same intersection point. Where there are several different intersection shotpoints for one intersection because of the number of horizons under consideration, two or three folds will normally suffice and frequently only one is needed.

Using this technique eliminates the need for guessed ties and gives the interpreter confidence in his or her work. It will be found that movements of 500 m (20 shotpoints on a typical marine survey) are perfectly normal, and deep horizons may be offset by as much as 2 km (80 shotpoints). Once the interpreter is satisfied that all the intersections tie accurately, the extra half day needed for the technique at the start of the interpretation is quickly recouped. As with many human activities, practice brings both improved speed and accuracy.

Contouring and mapping. At this stage the interpretation can be completed in the normal manner and the picked horizons can be digitized. Note that when the lines are posted on the shotpoint maps, the misties which are inherent in the data all reappear. To overcome this it is necessary to produce a "migration map" showing where the lines actually migrated compared with their positions recorded by the boat. Mark on a shotpoint map the intersection location for each horizon, do so for all the lines and color-code the points according to the picked horizon color. Each line will be found to have a set of positions which move it in an updip direction; these may be joined using a ruler, retaining the color-coding. (Figure 5 shows a typical end product for only one horizon for ease of visibility.) In most cases the dip varies across the area, and where no migration movement is observed, the ruled line runs through the true intersection. The pattern built up in this manner shows the area of maximum dip at a glance. Lines may move either way from an intersection depending on the dip direction.

Some common sense needs to be applied at this stage. The migrated positions of the lines are actually obtained from the lines which they cross, and it may sometimes be necessary to extrapolate the migrated position of an intersection up to a major fault or line-end if, for example, there are insufficient cross-lines to provide information. It will normally be more convenient to establish the migrated positions of the lines for all horizons on one map, and then copy the line positions on each individual posted value using a light table. Once the posted value map for a horizon is complete with the migrated positions of all the lines, contouring can commence.

When the data set consists of true dip and strike lines, the values should be projected orthogonally on the migrated line position, thus eliminating misties. The situation is slightly more complex when both lines are at an angle to the dip. Figure 6

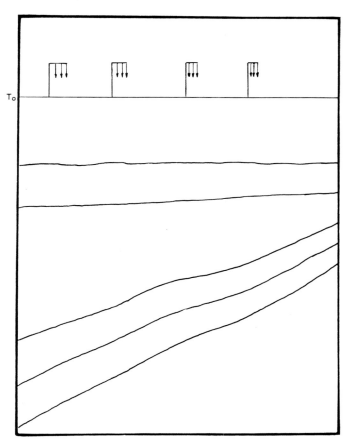

Figure 4. Final appearance of migrated stack.

shows the appearance of the migration map which would be observed under these circumstances. Data from point A now tie at point B, and data from point C tie at point D. The migration movement is indicated by the vectors AB and CD, and all posted values at intermediate points, such as E, will have to be projected to point F along a vector whose direction and length vary linearly along the line.

Fault patterns. It will be found that a major advantage of this technique is that fault patterns are simplified. Fault cuts should be marked on the migrated positions of the lines since they are affected as much as the values themselves. Consider the situation shown in Figure 7: AB is the geographical position of a line, $A_m B_m$ its migrated position, and CD a strike line. The fault is shown in its true position, and the actual seismic lines will show fault cuts with the throws indicated in Figure 8. This is because $A_m B_m$ represents not only the position of the line after the migration process, but also is the line which actually generates the data that are eventually recorded along AB. It can be seen that the faults can no longer be connected. This effect occurs whenever a fault plane runs between a line and its migrated position and is responsible for many interpretations which are overcomplicated and incorrect.

Accuracy of the method. In the ideal survey the grid will follow the true dip and strike directions, and the amount of migration observed will be the true migration. In the worst situation each line is at 45 degrees to the dip (see Figure 9). Each line here is subject to a dip of (true dip × cos 45) degrees, or about 70 percent of the true dip. It follows that the migrated line positions $A_m B_m$ and $C_m D_m$ have only shifted by about 70 percent of the amount that would have been observed had true dip lines been employed. In a survey in which the orientation of the lines varies randomly in relation to the dip, an average 90 percent migration correction is obtained.

In the situation shown in Figure 9 the angle θ of the vector $O - O_m$ indicates the amount of correction which has been achieved. When $\theta = 0$, $\cos \theta = 1$ and we have true dip and full correction. When $\theta = 45$ degrees, the correction is 70.7 percent. In this case the true location of the point O can be determined by extending $O - O_m$ by a factor of (100/70.7) to point O_t. Note that O_t is not directly accessible when the lines are not true dip and strike. This procedure will not normally be necessary but is available for those who require considerable accuracy of mapping.

An extra complication arises when more than one grid orientation has been employed. For instance, if some NE-SW lines are imposed on a generally NS-EW grid, then fluctuations of θ between the different data sets are likely to lead to conflicting posted values, particularly if the angled grid goes through one of the 90 degree intersections. This should be regarded as another warning to the interpreter rather than as a fault of the method. Strictly speaking, we could extend each of the vectors $O - O_m$ of Figure 9 by the appropriate amount for each line, thus overcoming the problem. Alternatively, the stacked sections could be interpreted and the final contour map subsequently migrated. This latter method could be employed in any case, but the interpretation resolution which is available on stacked data is poor compared with the use of migrated lines.

I have used the techniques described in this paper for the best part of a decade, and they have invariably succeeded. The confidence gained from mapping faults accurately and getting

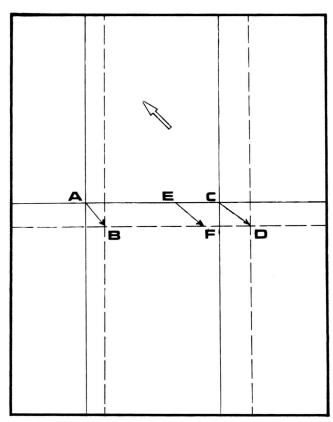

Figure 6. Treatment of migration map when both crossing lines are at an angle to the dip (see key on Figure 9).

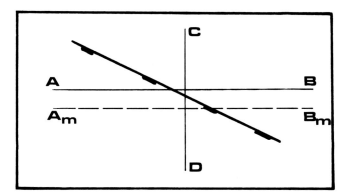

Figure 7. Faulting near an intersection in areas of dip.

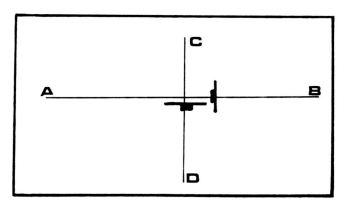

Figure 8. Fault cuts observed in Figure 7 situation.

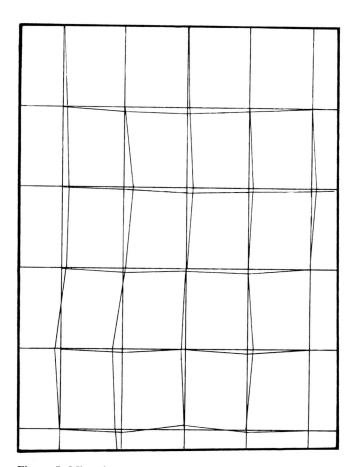

Figure 5. Migration map.

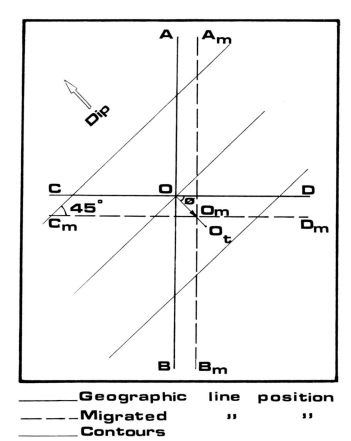

Figure 9. Intersecting lines at 45 degrees to dip.

accurate ties between lines is well worth the extra effort required at the onset of the interpretation. In the few cases where the intersections are demonstrated to be unattainable, it is advantageous to be aware of the problem at the start of the interpretation.

One final consideration arises from this approach. It is apparent that the lines which have data in the "wrong" place are the stacks: the horizon geometry has migrated the reflections out of their correct location. The so-called "migrated" lines have data in the correct position if they are true-dip lines, and the computer processing known as "migration" could more logically be called "de-migration." **LE**

(Acknowledgements: I should like to thank GECO a.s. of Stavanger, Norway, for permission to use their data, and in particular Paul Grogan for his valued comments.)

Colin F. O'Brien obtained an M.Sc. in Applied Geophysics from Imperial College, London, in 1966. He headed a large interpretation group at the Institute of Geological Sciences for some years before joining Cities Service in London. After further periods at Tricentrol and as chief geophysicist at Bow Valley in London, he became an independent consultant in 1984. Since that time he has worked on European and African acreage. He formed London Geophysics in April 1987, and in addition to interpretation services provides in-house interpretation courses. He is a member of the SEG, EAEG an PESGB.

Uses and abuses of seismic modeling

By SERPELL EDWARDS
Texaco, Inc.
Houston, Texas

Forward modeling, where an interpreter starts with a geologic model and generates synthetic seismic data (Figure 1), was first used by explorationists in the 1950s. Today, forward modeling packages have evolved to where they can handle complex 2-D and 3-D geologic models. Modern forward modeling programs yield synthetic field records, common midpoint gathers, stacked or migrated sections, vertical seismic profiles or 3-D seismic data sets. These flexible modeling programs have, in turn, induced an increasing number of explorationists to use them to answer questions related to acquisition, processing and interpretation of seismic data.

Flexible programs, when placed into the hands of interpreters unfamiliar with them, often translate into confusion and failure to correctly interpret and apply the results. Many problems encountered by interpreters while using modeling programs are independent of the software and can be avoided. This paper focuses on some major uses of forward modeling programs and a few of the problems interpreters have with them.

Model input. One of the critical factors determining the type of seismic modeling that can be performed is whether the geologic model is defined in one, two, two and one-half, or three dimensions.

A 1-D geologic model implies that the model is a one-sample-point-wide description of the geologic column at one location on the surface of the earth. A geologic model derived from a suite of logs from one well represents a 1-D model (Figure 2). Most seismic modeling programs assume that a 1-D geologic model can be expanded into a "flat-earth" model where all the bedding planes are flat, horizontal surfaces and each rock unit is homogenous. During program execution, algorithms use these assumptions to introduce 2-D or 3-D earth effects. One-D models are often used to:

- Derive synthetic seismograms from well logs
- Align well logs with the seismic sections
- Extract seismic wavelets
- Generate synthetic, zero-offset vertical seismic profiles (Figure 2)

Two-dimensional models represent a vertical slice of the earth. The geologic cross section is a good example of a 2-D geologic model (Figure 2). Two-dimensional modeling programs usually present their results in a manner that mimics the more traditional seismic sections with which interpreters are familiar.

Typically, 2-D modeling packages generate:

- Migrated and unmigrated time sections from depth models (Figure 5) which are often used to verify interpreters' picks
- Displays showing the effects of oil and gas saturation
- Two-dimensional field records, CMP gathers, and stacked data sets to evaluate new processing algorithms
- Synthetic field records or gathers to show the effects of offset on amplitude and phase (Figure 6)

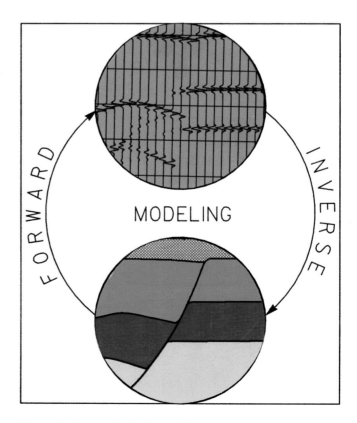

Figure 1. Types of seismic modeling.

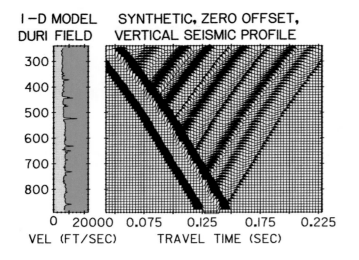
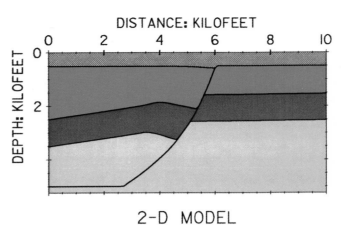
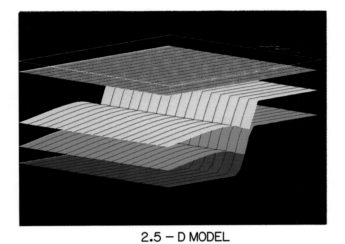

Figure 2. Examples of 1-D, 2-D, 2.5-D, and 3-D geologic models.

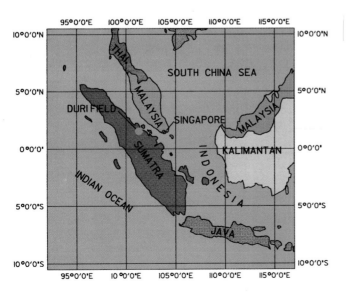

Figure 3. Location map of the Duri field, Sumatra, Indonesia.

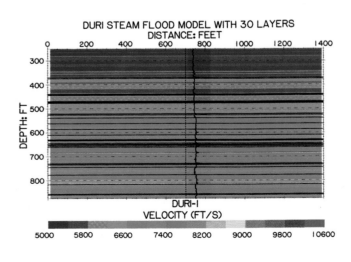

Figure 4. Thirty-layer, 2-D velocity model from Duri field.

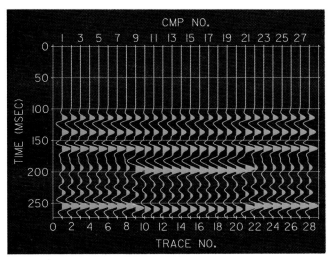

Figure 5. Synthetic seismic section from Duri field, showing the effects of steam injection. The central portion of layers 15 and 16 have had their velocities reduced by 30 percent.

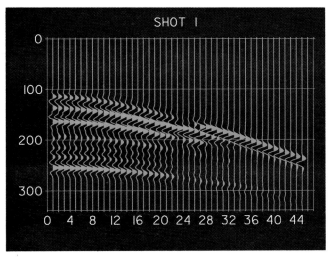

Figure 6. Synthetic field record from the Duri model, showing the effects of the shallow, high-velocity carbonate layers.

Recently, the term *two-and-one-half dimensional models,* or *2.5-D* has been accepted by the industry. These are 2-D models that have been extended into the third dimension by adding, in the strike direction, copies of the original section (Figure 2). In 2.5-D models, cross sections perpendicular to the original 2-D line will appear to contain only flat, level horizons. Two and one-half dimensional models are valuable since they permit the use of 3-D seismic modeling algorithms on 2-D data.

In 3-D models, all depth or time values that can define a surface may be different. Most 3-D geologic models can be thought of as a set of structure maps with each horizon correctly aligned and stacked on top of the next deeper horizon. Figure 2 contains an example of a 3-D model of a rollover anticline. By taking into account the third dimension, 3-D algorithms can better model the effects of the earth on the propagation of seismic energy. Some of the advantages of 3-D modeling are:

• Handling of "out of the plane of section" problems
• Generation of crooked-line solutions
• Testing, prior to going to the field, of the alignment of new seismic lines over a prospect
• Generation of test data sets and binning analysis prior to shooting 3-D surveys
• Creation of synthetic, slant-hole, walk-away VSP surveys
• Creation of 3-D displays that help sell prospects to management or clients

Regardless of the model's complexity, seismic modeling programs require impedance changes within the model in order to produce seismic events. The major variable controlling impedance changes within rock units is usually the velocity change

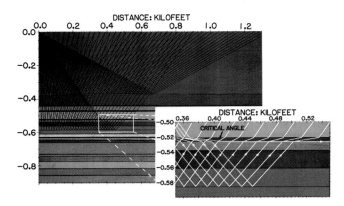

Figure 7. Ray-tracing plot off layers 1 and 15 of the 30-layer Duri model.

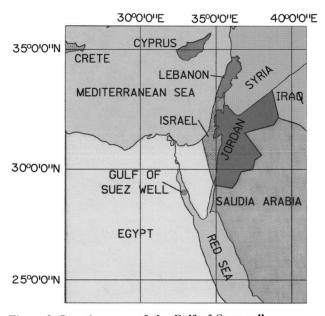

Figure 8. Location map of the Gulf of Suez well.

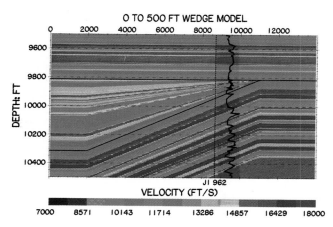

Figure 9. Gulf of Suez 2-D model made from a blocked velocity log.

encountered when going from one lithology to another. Rock velocities can be obtained from sonic, velocity or VSP well logs, or from the processing of seismic data. Density information comes from well logs, geologic studies, or it may be derived from some velocity-density relationship.

Seismic modeling programs that produce more than just pressure-wave solutions usually require additional input; i.e., Poisson's ratios, P-wave to S-wave ratios, P-wave and S-wave quality factors and vertical velocity gradients.

Modeling techniques. Seismic modeling is like any other scientific investigation. The first step is to define the problem. The objectives should be stated, including the type of displays or computer files that will be needed when the project is finished. All too often the interpreters have not thought through their problems and established objectives. Their only reason for modeling is to answer "yes" to their manager's or client's question, "Have you modeled this?" An example of a problem is *How can we see through this high-velocity layer?* The objectives could be to test certain shot-and-receiver geometries with the output being synthetic field records with plots sent to a hardcopy device and a standard SEG Y tape sent to the processing center.

The interpreter should check to make sure there are enough data available to build a model capable of doing what is desired.

One of the recurring problems that new users have is that they want to generate realistic synthetic seismic sections from 3-D geologic models in structurally complex areas, yet they have only mapped the zone of interest.

Once the objectives are known and the modeling data assembled, the program must be matched to the job. There is no substitute for thoroughly knowing how a program works. This does not mean you have to understand the algorithms, but the user should know what the program capabilities are and how to manipulate the program to produce the desired type of results. No program should be expected to do its best if the user refuses to read the manual or help screens and has not taken time for hands-on training prior to running a complex problem.

One common abuse of modeling software is the overkill syndrome. Inexperienced users often pick the most complex and CPU-intensive programs to solve relatively simple problems. Generally, the fewer dimensions in a model, the easier a program is to use and the quicker it runs on the computer. One-dimensional models usually execute within seconds while 3-D models with full elastic solutions may take days to complete. The solution to the overkill syndrome is to define the problem, list the objectives, and then match the problem to a modeling program.

There is no substitute for knowing how a program works. Interpreters that do not have time to learn a program should obtain assistance from a knowledgeable user. That data for the geologic model, a brief description of the problem, and a list of objectives should be taken to an experienced user and — as a team — the problem should be set up and executed.

As in all scientific experiments, the experimenter should document what is done while it is being done. Nothing is more frustrating than executing a program a dozen or more times and then not being sure of the variables used for each of the runs. Machine files, screen dumps, or plots should be made to document the program configuration, program variables, and input data values. It is always better to have made too many screen dumps and plots than it is to finish a model only to have to rerun it in order to verify the parameters used to produce all of the needed plots.

Case histories. Even with a well defined problem and with specific objectives, the interpreter needs to be actively involved with the execution of the program and the analysis of the results. Should the program yield unexpected results, the interpreter may have to rethink the problem and the objectives.

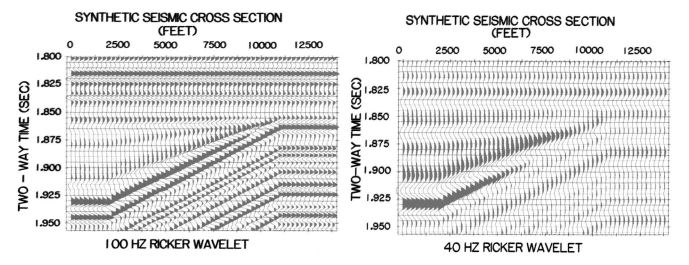

Figure 10. Zero-offset, migrated, synthetic seismic sections derived from the Gulf of Suez expanding sand-wedge model. Note the apparent onlapping seismic signature in the 40 Hz synthetic section.

History 1: Duri field model, Sumatra, Indonesia.

In 1985, a project was undertaken to determine if a seismic survey in the Duri field could detect the propagation of a steam front through the reservoir sands (Figure 3). The objectives were to measure:

• Time shifts caused by the reduction in velocity as steam invaded the reservoir

• Changes in reflection amplitude on stacked sections

• Changes in amplitude and phase with offset on CDP gathers

The Duri field is very shallow, with the zone of interest at this particular injection well roughly 600 ft (182.88 m) beneath the surface. Initially, a 3-D model was made by blocking the sonic log into 11 flat, homogenous layers with five layers above and five beneath the reservoir sand. Instead of using this large, time-consuming 3-D model, we chose a 2-D modeling solution since no significant faults were present and the engineers were willing to assume a symmetrically advancing steam front. A 1-D "flat earth" model wasn't used because a horizontally varying velocity was needed in the injection sand.

After analysis of the sonic log by an interpreter familiar with the field and the modeling software, it was decided that the original model was not sufficiently detailed to accurately model the reservoir sands. The number of horizons was increased to 30 (Figure 4). For the pre-steam model, the velocities were taken directly from the sonic log. The velocities in the center of the reservoir sands (layers 15 and 16) were then reduced by 1 percent, 5 percent, 10 percent, 20 percent, 30 percent and 40 percent to show the effect of the steam. Figure 5 indicates that a 30 percent reduction in velocity should produce an apparent bright spot plus a sag in the underlying reflections.

Figure 6 is a simulated field record from the Duri model. One of the unexpected results of this modeling effort was the lack of any reflected energy from the zone of interest being picked up by any geophones located more than 900 ft (274.3 m) from the source. Upon analysis of the ray-tracing plots (Figure 7), it became apparent that shallow, high-velocity carbonate layers above the reservoir sands were refracting the *P*-waves at critical angle, effectively masking the deeper horizons. Not only did this show that field tests needed to be run to determine maximum geophone offset but that amplitude with offset studies would probably be useless.

History 2: Gulf of Suez model

Another project used sonic log data from a Gulf of Suez well (Figures 8 and 9) to model the seismic response of a reservoir sand. The objective was to show the effects of various pay-sand thicknesses on the seismic sections. A 2-D model was designed where the pay sand varied in thickness from 0-500 ft (0-152.4 m) with the wedge-shaped expansion spread evenly over the entire zone. The sediments above and below the pay sands were not changed.

Figure 10 shows the response of the Gulf of Suez model at 100 and 40 Hz. Both of these figures help to answer the original question *What will the seismic response be at various frequencies and at various thicknesses?* What was not anticipated was the seismic stratigraphic implications of the modeling. The 100 Hz synthetic seismic section clearly shows the expanding wedge of sediments while the 40 Hz synthetic shows the onlapping reflectors.

Over the next several years, the art of interpretation will become more scientific. Interpreters will shift their skills from pushing color pencils and folding paper sections to using interactive computer workstations where digital earth models are the norm. Prior to drilling, forward modeling will be done to lay out new seismic lines needed to firm up a prospect, to confirm the results of special seismic processing and to verify geologic concepts.

By remembering that seismic modeling is a scientific experiment and by following sound procedures for conducting scientific research, new users will encounter fewer problems and have greater confidence in their results. **LE**

G. Serpell Edwards is a research associate at Texaco's geophysical research facilities in Houston. He received a B.S. in geology from Tulane University and a Ph.D. in oceanography from Texas A&M University. He worked as an oceanographer in Texaco's Research Lab during 1971-79. From 1979-83, Edwards lived in Sumatra where he worked as an interpreter with Caltex Pacific Indonesia. Since 1983, he has been involved with seismic interpretation and modeling research at Texaco.

Complex seismic trace attributes

By JAMES D. ROBERTSON and DAVID A. FISHER
ARCO *Oil and Gas*
Dallas, Texas

Interpretation of seismic reflection records is a blend of quantitative science and visual art. An interpreter uses the rules of wave propagation, realistic limits on impedances of rocks and accepted principles of stratigraphic deposition and structural deformation to arrive at a geologic understanding of a seismic section. This application of science, however, seldom produces a unique answer, and the obligation to consider alternative interpretations provides plenty of latitude for creative imagining.

Whether making measurements or conjuring hypotheses, the seismic interpreter is heavily influenced by the quality and character of the seismic displays. What we see, or think we see, on the records sets off a string of assertions about trapping, reservoir quality, hydrocarbon charging and the like that become the raw data for prospect economic evaluations and drilling decisions. Some aspects of a typical interpretation are obvious and invariant if the seismic section is colored, squeezed, enlarged, gain-adjusted, etc. Other aspects are subtle and can be reinforced, altered or even obliterated by a change in the way the information is exhibited.

This contribution to the series reviews a class of seismic displays called *complex seismic trace attributes*. The standard seismic section is a plot of particle velocity at the geophone (or acoustic pressure at the hydrophone) versus time. At any particular point in time, this measurement depends on both the amplitudes and phases of the individual frequencies that comprise the seismic disturbance. Computing complex seismic trace attributes basically is a transformation that splits apart the amplitude and angular (phase and frequency) information into separate displays. The information in the seismic section is mathematically manipulated to produce new displays that highlight amplitude or angle at the expense of eliminating the other. The word *complex* is a part of the name of the manipulation not because the procedure and output are complicated, but because the computation assumes that the conventional seismic trace is the real part of a complex mathematical function. (The imaginary part is the Hilbert transform of the real part.) No fundamentally new information is created by the manipulation; rather, alternative sections are produced that sometimes point out aspects of the geology that were masked on the conventional section.

Seismic attributes are the offspring of an approach to signal analysis originally developed by D. Gabor and others in the mid-1940s. Gabor noted that signals, such as frequency-modulated radio transmissions, are not amenable to conventional Fourier analysis because their frequency content varies with time. Fourier analysis decomposes a signal into its sinusoidal components based on the assumption that frequency content is not changing with time; in electrical engineering parlance, the signal is "stationary." In the Fourier approach, a signal thus can be represented either by its time description or by its Fourier phase and amplitude spectra. Geophysicists can appreciate Gabor's discomfort with applying Fourier analysis to his radio waves because we know from our common experience that the spectral content of our seismic traces varies with traveltime through the earth; typically, high frequencies are preferentially lost. We usually compensate for this physical reality by estimating Fourier spectral content over short windows — rather than for the full trace length — and by applying deconvolution and other spectral alteration algorithms in a time-varying manner. Gabor made the more generalized assumption that a signal can be a function of both time and frequency, i.e., it is "nonstationary." He then proceeded to borrow a mathematical formalism from quantum mechanics to describe such a signal. That description is now known as *complex trace analysis*.

Complex trace analysis and the associated computations of attribute displays were picked up and applied to seismic sections in the early 1970s by Nigel Anstey and others at Seiscom Delta. The technique was publicized in AAPG and SEG Distinguished Lectures by Turhan Taner and Robert Sheriff, respectively, in the mid-1970s and was the subject of papers published by these two individuals and Fulton Koehler in AAPG *Memoir* 26 (1977) and in GEOPHYSICS (1979).

Attribute displays. Complex seismic trace analysis produces sections known as *instantaneous attribute displays* because the attributes are computed on a sample-by-sample basis. Amplitude information is encoded in the envelope amplitude attribute, also called *reflection strength* or *instantaneous amplitude*, which is a robust, smoothed, polarity-independent measure of the energy in the trace at a given time. Envelope amplitude plotted in color highlights bright spots, amplitude variations caused by thin-bed tuning, major lithologic changes and general variations in reflectivity in a format that promotes visual equality between peaks and troughs at the expense of removing polarity information.

Angular information is generally displayed as instantaneous phase or instantaneous frequency (the time derivative of instantaneous phase). Instantaneous phase is the phase at an instant along the trace — as distinct from the phase shift over a window which Fourier analysis determines. Instantaneous frequency is a sample-by-sample estimate of the trace's dominant frequency — as distinct from the average frequency over a window. The two attributes conventionally are computed and plotted in color at a time sample without consideration for the envelope amplitude at that time sample. In practice, the two angular attributes are most often employed to display that ambiguous entity known as *seismic character*. In this mode, the attributes are empirically useful for tracing bed continuity, picking event terminations and unconformity surfaces in a seismic stratigraphy analysis, and mapping wavelet variations and thin-bed interference patterns that have been geologically calibrated by well control.

A. W. Rihaczek in 1968 and M. H. Ackroyd in 1970 pointed out in the electrical engineering literature that instantaneous frequency is a measure of the center frequency corresponding to the first moment (i.e., the mean) of the complex energy density function (i.e., power) of the signal at a given time. What this means to geophysicists is that the instantaneous frequency at the peak of a zero-phase seismic wavelet is equal to the average frequency of the wavelet's amplitude spectrum (see *Complex seismic trace analysis of thin beds,* Robertson and Nogami, GEOPHYSICS, April 1984). Thus, there are points on a conventional instantaneous frequency trace where instantaneous frequency directly measures a property of the Fourier spectrum of the wavelet.

Instantaneous phase likewise measures the wavelet's true phase at these same points. Unfortunately, the physically meaningful measurements occur at only a small percentage of the trace's samples and are obscured by adjacent (in time) instantaneous taneous phase and frequency at the maxima, and assigning these phase and frequency values to all the time samples between the adjacent amplitude minima. The results are blocky traces that highlight phase and frequency in a physically more meaningful manner than conventional attribute displays. Response attributes do not attempt to estimate phase and frequency at time samples away from energy maxima where the estimates lose physical significance. It is important to recognize that there is a tradeoff associated with response attributes. Vertical resolution is reduced compared to instantaneous attributes or the conventional section, and the fine detail of the character of an anomaly may be lost.

Wedge model. The appearance and utility of attribute displays can be demonstrated by computing the attributes of simple seismic models. A low-impedance wedge in a homogeneous halfspace is shown in Figure 1. The model could represent the pinchout of a particular lithology in a lateral facies change, the thinning of a gas-saturated zone in a sandstone reservoir, a variation in porosity in a carbonate or a number of other geologic situations of exploration interest. The actual velocities and densities assigned to this particular model are appropriate for the pinchout of a Miocene hydrocarbon reservoir buried at moderate depth in the offshore Gulf of Mexico.

The synthetic seismogram of the model is represented in Figure 2a. The thickness of the wedge is denoted in units of the dominant period, T, corresponding to the dominant frequency of the

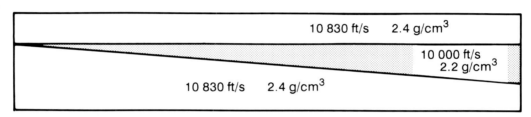

Figure 1. Geometry and seismic parameters of wedge model.

attribute values that often have lateral (trace-to-trace) continuity but do not correlate with significant Fourier properties. The mix of meaningful and meaningless values is probably the major factor that has frustrated interpreters looking for physical significance in the actual numbers on attribute sections. The mix has produced a commonly held belief among interpreters that attribute sections are nice to look at and sometimes empirically useful for mapping character, but difficult to use when the task is to derive quantitative parameters that can improve chance factors on reservoir quality, trap size, pore fluid type and the like.

An important advance in attribute analysis occurred in 1986 when J. H. Bodine published an article in the *Oil and Gas Journal* that directly addressed the issue of the physical meaning of instantaneous phase and frequency sections. Bodine pointed out that significant energy bursts (i.e., reflections) on a seismic trace can be defined as the envelope pieces located between successive minima on the envelope amplitude representation of the trace. A local maximum obviously exists between any two minima, and one can selectively extract and plot the instantaneous phase and frequency associated with these maxima. Since most of the signal energy in a trace is found in the vicinity of these peaks, the angular properties of the waveform defined by each envelope piece can be accurately described by calculating the attributes only at the peaks. Instantaneous phase tends to be linear around the amplitude maxima, and the instantaneous frequency at a maximum, as noted above, measures the average frequency of the amplitude spectrum of the waveform. Bodine gave the names *response phase* and *response frequency* to the proposed attribute representations.

Response attributes are computed and displayed by locating envelope amplitude minima and maxima, computing instan-

spectrum of the zero-phase wavelet used in the modeling. Where the wedge is thick, the top and bottom are marked by the central lobes of the corresponding reflections. The polarity convention is that a negative reflection coefficient like the top of the wedge is a trough; a positive reflection coefficient like the bottom is a blackened peak. The reflections interfere with each other as the wedge thins, and the top and bottom are no longer resolved by differential event time below about $T/2$. The point, $T/2$, corresponds to the quarter-wavelength tuning point where amplitude reaches a maximum value owing to constructive interference before falling off monotonically as the bed thins to zero thickness and the top and bottom reflections destructively interfere.

Envelope amplitude, instantaneous phase and instantaneous frequency are displayed in Figures 2b, 2c and 2e. Response phase and frequency are displayed in Figures 2d and 2f. On each attribute section, dotted lines that track the main upper trough and main lower peak on the conventional section are used as locators to reference the color displays to each other and to the original seismic data. The amplitude tuning at a half-period stands out in red on the amplitude display, and the downward progression through the color bar as the wedge thins to zero clearly show the decrease in amplitude caused by destructive interference. The various interference patterns associated with the wavelets and their sidelobes stand out on the instantaneous phase and frequency sections. Of particular interest is the anomalous increase in instantaneous frequency that appears when the wedge thins below $T/4$. This increase is highlighted by the color change from blue to red as the wedge thins to zero. We know that a thin, low-impedance bed differentiates a wavelet, thereby raising the average frequency of the composite reflection relative to the basic wavelet. The instantaneous frequency increase

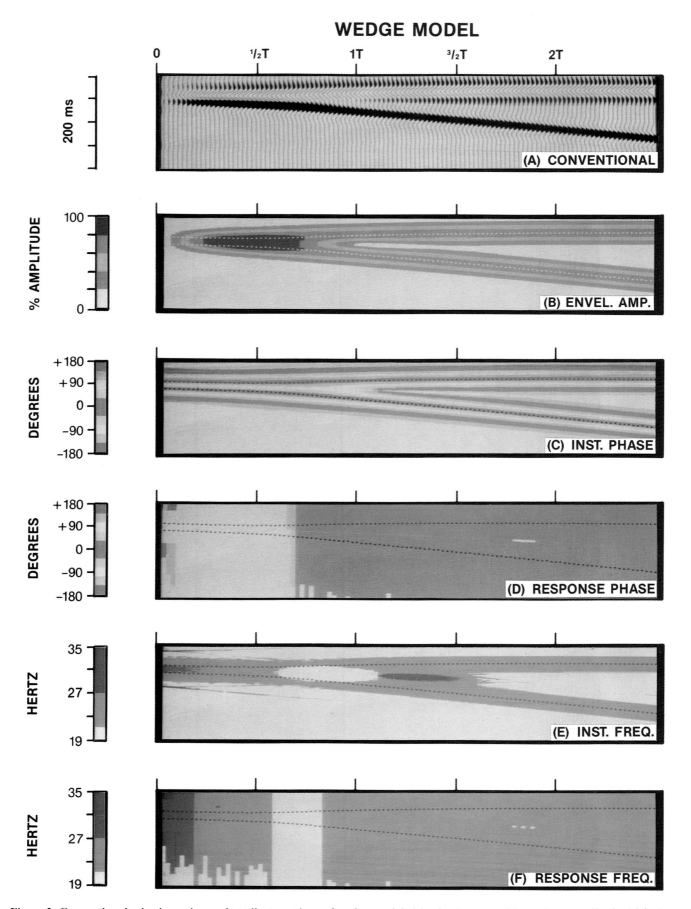

Figure 2. Conventional seismic section and attribute sections of wedge model: (a) seismic traces, (b) envelope amplitude, (c) instantaneous phase, (d) response phase, (e) instantaneous frequency and (f) response frequency.

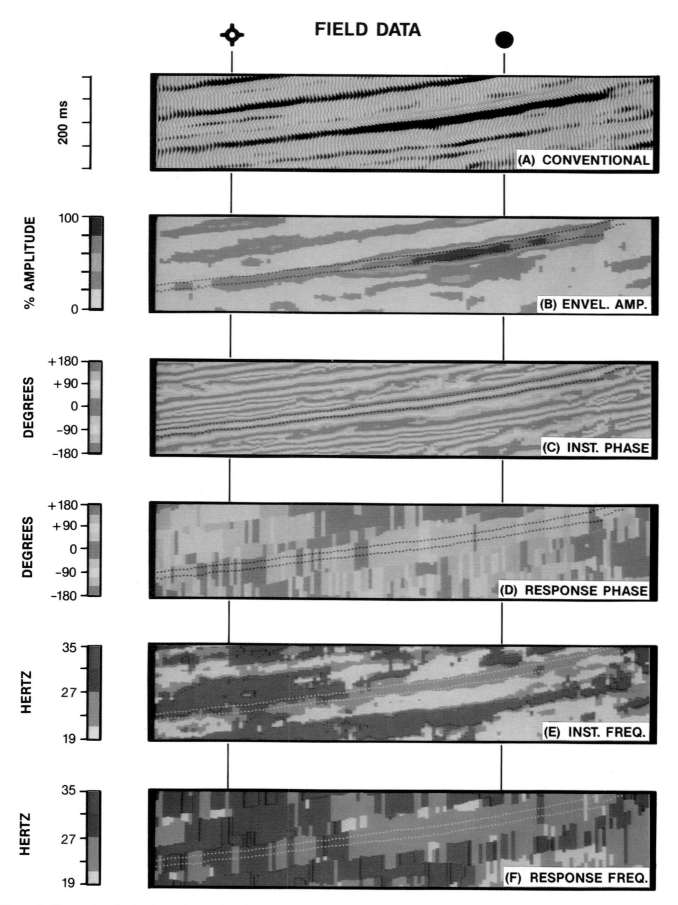

Figure 3. Conventional seismic section and attribute sections of field data: (a) seismic traces, (b) envelope amplitude, (c) instantaneous phase, (d) response phase, (e) instantaneous frequency and (f) response frequency.

demonstrates this phenomenon and is characteristic of tuning at small fractions of a wavelength. The increase can be used to track and map the feather edges of low-impedance thin beds.

The response attributes highlight the instantaneous phase and frequency values associated with amplitude maxima. Where the wedge is thick, the response phase (Figure 2d) verifies the zero-phaseness of the basic wavelet (either 0 or 180 degrees), and the response frequency (Figure 2f) is a quantitative measure of the average frequency (22.5 Hz) of the spectrum of the model wavelet (a 20 Hz Ricker). As the wedge thins, the response attributes map and calibrate the angular changes associated with the thin-bed tuning. Response phase changes to 90 degrees between T and $T/2$, signifying that the thinning of the bed has differentiated the basic wavelet. Response frequency, like instantaneous frequency, steps up the color bar into the red color as the wedge goes from $T/4$ to zero.

It is important to note that the various attribute sections change minimally between about $T/4$ and $3T/4$ at the levels of the binning used for the various color bars. The binning levels approximate what is generally required to smooth out noise in real data. Since the $T/4$ to $3T/4$ zone corresponds to frequently encountered reservoir thicknesses, the modeling implies that attribute sections are more robust at defining reservoir edges than mapping variations within the body of reservoir.

Field data. Figures 3a to 3f are the conventional true-amplitude section and five attribute displays of a portion of a seismic line from the offshore Gulf of Mexico. The high amplitude event in the right-center of the section is a thin (less than $T/2$ thickness), hydrocarbon-bearing Miocene sand reservoir that is a real life analog to the wedge model in Figure 2. The acoustic response of the sand is controlled by the presence of the hydrocarbons, so the high amplitude event is a proven bright spot. A 30-ft pay zone, corresponding to a bed thickness of slightly less than $T/4$, is present in the producing well on the right. The pay zone is absent in the dry hole on the left and presumably pinches out between the two wells.

The attribute displays reinforce and add detail to the interpretation made from the conventional seismic section and borehole information. As in Figure 2, dotted lines are used in Figure 3 to track the trough and peak bounding the zone of interest in order to relate the color displays to each other and to the seismic data. Envelope amplitude appears to be stronger about 10 traces downdip from the producer than at the wellbore itself, suggesting that the pay zone is closer to $T/2$ tuning and hence thicker there than at the well. Some independent evidence supports this inference in that a third well about 1000 ft along strike from this amplitude buildup contains over 50 ft of pay. Farther downdip, envelope amplitude progressively diminishes, implying that the reservoir is pinching out toward the dry hole.

The zone of interest has a uniform appearance all across the section on the instantaneous phase display (Figure 3c). The top of the zone is marked by a continuous 180 degree event (blue color) and the bottom by a continuous 0 degree event (orange color). This character mirrors the character of the phase below $T/2$ in the wedge model, again pointing out that a thin bed differentiates a zero-phase wavelet. Response phase (Figure 3d) is reasonably constant and a few tens of degrees off 90 degrees where instantaneous amplitude is high, implying that the impedances in this real data example are somewhat different above and below the zone of interest, so the wavelet differentiation is not quite symmetrical. Where amplitude diminishes downdip, response phase becomes quite variable. The implication is that the character of the reservoir is changing, but the exact physical significance, if any, of the variability in response phase is not clear.

Instantaneous frequency and response frequency (Figures 3e and 3f) maintain values between 21-27 Hz (blue color) from the producer downdip to about 20 traces from the dry hole. At that point frequency rises above 27 Hz (the red color appears) over significant lateral distances. By analogy with the wedge model, the onset of the color change can be interpreted as the location where reservoir thickness becomes much less than $T/4$ — somewhere between zero and $T/10$ based on the modeling. A reasonable posture for the explorationist might be to equate the onset of the red color in the bed with the downdip edge of the producible reservoir and to map and make reserve estimates accordingly until new well control and production information provide more information.

Also of note in Figure 3 is an anomalous increased separation on the instantaneous phase display between the orange zero-phase event at the base of the gas reservoir and the next orange event below the base. This increased separation appears to be a low-frequency shadow. Such shadows sometimes correlate with and are interpreted to be direct indicators of overlying hydrocarbons. As is true in this case, attribute sections can highlight the shadows. The phenomenon is easier to see on the instantaneous phase section than on the instantaneous frequency section owing to the selection and binning of the colors.

Complex seismic attribute displays can enhance the interpretation of a conventional seismic section both by highlighting character changes in the data and by assisting with quantitative estimates of wavelet characteristics and stratigraphic variables. Since the various attributes emphasize different properties of the seismic signal, the most profitable approach generally is to produce the entire suite of attributes, calibrate the sections with well control and then draw inferences with a cautious appreciation of the benefits and the pitfalls associated with each attribute. The mechanics of producing the attribute sections are becoming progressively easier with the advent and evolution of color graphics workstations and color plotters, so the incremental cost of this interpretive technique is steadily decreasing. **LE**

(Acknowledgements: We thank Robert Sheriff for helpful comments that improved this article and ARCO *Oil and Gas for permission to publish it.)*

David A. Fisher is a research geophysicist for ARCO *in Plano, Texas. He holds a B.A. degree in geology from the University of California at Santa Barbara and an M.S. in geophysics from the University of Houston. During 1980-81 he worked as a processing geophysicist for Geophysical Systems in Pasadena, California. He has worked for* ARCO *since 1983, primarily on interactive computer techniques to assist the interpretation of seismic data.*

James D. Robertson is manager of ARCO *Oil and Gas Company's exploration staff in Dallas, Texas. He holds a B.S.E. degree in civil and geological engineering from Princeton University and a Ph.D. in geophysics from the University of Wisconsin. He is a member of SEG, AAPG, SPE and EAEG, a past president of the Dallas Geophysical Society, a former lecturer in SEG's Continuing Education Program and a current member of the Editorial Board of* The Leading Edge. *He received SEG's Outstanding Presentation Award in 1980 and 1984, and SEG's Outstanding Paper in* Geophysics *Award in 1986.*

Balanced cross-sections—
an aid to structural interpretation

By PETER B. JONES
Thrustbelt Systems
Calgary, Canada

Some of the last major reserves of oil and gas await discovery in the fold and thrust belts flanking the world's sedimentary basins. Historically, exploration for petroleum began in these deformed belts, topographically expressed as foothills, where structural traps could be located by geologic mapping. The obvious folds were drilled at an early stage of exploration, but deeper drilling often showed the underlying structure to be complex and highly faulted. Early seismic exploration was incapable of resolving these complexities. Now, many deformed belts are being reexamined with improved techniques and a more systematic application of structural geologic principles.

The most important principle is that layered sedimentary rocks are folded and faulted by a process of "thin-skinned" tectonics in which there is a lower limit to the deformation—a detachment zone above which the rocks are compressed horizontally, creating folded and faulted hydrocarbon traps. Locally, the basal detachment may be a shale or evaporite; on a crustal shale, it may be the Moho. The other factor is the strong influence of layering or bedding planes on the geometry of both normal and thrust faults. Normal faults move down-section in their direction of movement or along bedding planes, while thrust faults climb upsection or follow bedding planes.

Deciphering the subsurface structure of deformed belts requires the construction of structural cross-sections perpendicular to the structural strike, based on geology, well information, seismic profiles and other data. Wherever possible, these cross-sections should be retrodeformable or "balanced." A balanced cross-section represents a possible consequence of faulting and folding of an undisturbed rock sequence to form a linear fold or thrust belt without addition or removal of material. This means several things:
• The cross-sectional area of a profile through that thrust or fold belt should have the same area as that of the original sequence of rocks (less erosion, of course).
• The offset of any fault is constant through all the layers it cuts across or is split between a number of smaller faults.
• The thickness of any bed after deformation equals its original thickness.
• When each faulted element is moved back along the fault plane by the amount of offset shown, it will fit back to its place of origin.

This process of checking a cross-section has been described as "filling the container."

Computer modeling shows that some styles of structural interpretation actually violate both geometric and basic geologic principles.

Interpreted seismic profiles are a common form of cross-section. Two reflectors, juxtaposed by thrusting, can easily give a spurious reflection character that can be mistaken for some other unit, and many dry holes have been drilled as a result. Balancing the interpreted seismic profile can often avoid such expensive errors by showing that it is not possible to move the section along the fault planes shown and still match the final interpretation. Although a balanced cross-section is not necessarily correct, an unbalanced one is always wrong.

Construction of a balanced cross-section is a tedious process that may take weeks. Even a small alteration to one part of a section means redrawing the whole. Nobody who has spent a month in this type of operation has time left for testing an alternative interpretation or incorporating new data properly by repeating the whole process. Many cross-sections, though painstakingly constructed and balanced, are impossible interpretations because, although the deformed state can be matched to an undeformed state, the intermediate stages cannot be matched to either.

Although tedious, balancing a cross-section is basically a

simple chore, and the principles involved can be incorporated in computer programs that simulate the movement of normal and thrust faults rigorously and objectively. They can also maintain a constant cross-sectional area throughout fault movement, ensuring that the resultant cross-section is restorable. Most folds, which are the products of fault movement, can also be modeled. Even folds formed by evaporite and shale flowage, which result from displacement along an infinite number of minute faults, can often be represented as the product of a finite number of large faults. Detachment zones, above and below which rocks of differing competence behave in different ways, can be represented as bedding plane faults.

There are two types of computer programs for modeling geologic structures: restoration and forward modeling. The restoration method takes a cross-section that has been drawn to fit the data points—surface geology, well tops or a seismic profile—and attempts to move the material back to fit an assumed model of the original, undisturbed stratigraphic sequence. If it does not fit (i.e., is not balanced), adjustments must be made to the interpreted cross-section or seismic profile and the process repeated until the profile can be fitted back. The alternative approach, forward modeling, begins with the assumed undisturbed section, inserts faults in the appropriate (or estimated) positions and amounts of slip, and checks the computer-drawn balanced cross-section to see if it matches the actual data points. If it doesn't, this process also must be repeated using revised parameters. Complex structures are modeled in stages, one or two faults at a time.

Intuitively, the former seems the logical approach, but both methods involve similar amounts of trial and error. Unless a hand-drawn cross-section has been carefully constructed, probably requiring weeks of work, it will not restore back to a smooth, undisturbed profile. Because of the complex interactions between adjacent faults, the reasons it doesn't restore are not necessarily obvious nor easily corrected. Since there is always erosion of material after subaerial (as opposed to submarine) deformation, a restored cross-section contains large blank areas. In forward modeling, structures above the present erosion level are included in the model. This can give an insight into the physiography before erosion, as well as the structure along the strike and down the plunge. Regardless of the system employed, the user must have a solid grounding in structural geologic principles.

In forward modeling, the computer simulates the progressive movement along fault planes, so that each balanced cross-section

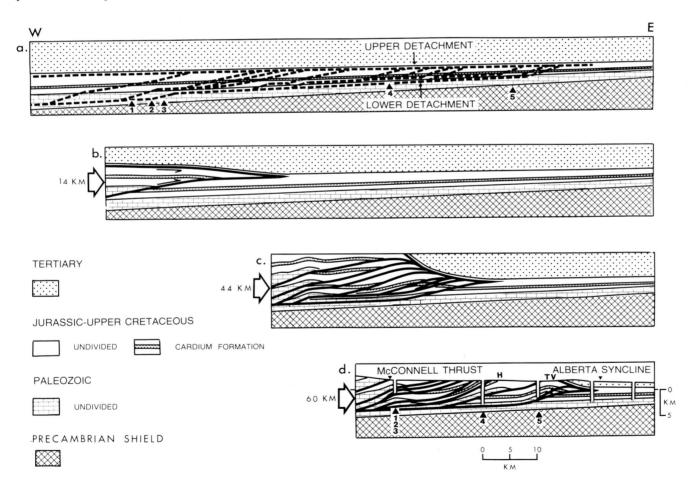

Figure 1. Evolutionary stages of southern Alberta foothills: (a) Undisturbed sedimentary sequence—dashed lines show positions of future thrust faults, as well as upper and lower detachment zones. (b) After movement of first thrust fault. (c) After movement of ninth thrust. (d) Present structure resulting from movement of 17 major thrust faults, showing Highwood structure (H) and Turner Valley oil field (TV). Traced from printout of Thrustbelt model. Numbers 1, 2, 3, 4 and 5 show well penetrations of the Mississippian in their present (stage d) and restored (stage a) locations. All cross-sections are balanced with respect to each other and to the undisturbed profile. (The original cross-sections were modeled on a scale of 1:48 000, using an XT clone, and are up to 12 ft long. Eleven computer runs were required to produce a satisfactory fit with the well data and surface geology.)

contains an infinite number of balanced cross-sections of intermediate stages of deformation. This capability was recently used to create large numbers of frames, each one a balanced cross-section, for the animated sequences in *Birth of the Rockies*, an educational film made by the British Broadcasting Corporation.

The two most important exploration applications of computer modeling are the subsurface interpretation of deformed belt structures from well, seismic and surface geologic data (Figure 1d) and the construction of "palinspastic" maps of different stratigraphic units. A palinspastic map (Figure 2b) shows the occurrences of a given rock unit in wells and outcrops restored to the locations where they were deposited, before faulting and folding transported them to their present locations. It is not possible to determine porosity trends and facies boundaries accurately in deformed terranes until such a restoration has been carried out. Forward modeling automatically provides a palinspastic (restored) cross-section (Figure 1a) corresponding to each completed cross-section (Figure 1d). In once recent study, the two functions were combined to track a linear porosity trend that ran subparallel to the structural strike and then to determine its position within a stack of folded thrust sheets, resulting in a dual gas discovery, using the method shown in Figure 2. In another case, computer modeling proved that a particular overthrust play was impossible. The thrust sheet that the client needed to make the play work wouldn't fit in a balanced structural system.

As with any tool, there is a learning period, and the user must possess some experience in geologic and geophysical interpretation in deformed belts to make proper use of computer modeling. The crust of the earth is a heterogeneous body, and it is not possible to make detailed geometric predictions of progressive deformation. Given a reasonable fit with the data, there comes a point where it is not worth the time involved to try to match the fine detail that may result from very small, local and unpredictable inhomogeneities. This can be done quickly by hand, confirming or eliminating an interpretation long before this stage is reached. Even so, it is amazing how closely an actual structure can resemble a model made with simple and rigorous geometric and geologic constraints, and virtually all fold and thrust belt structures can be simulated.

Computer modeling shows that some styles of structural interpretation actually violate both geometric and basic geologic principles. The most common example is the seismic profile in which a fault is interpreted to follow a narrow, near-vertical zone of poor data, often above the leading edge of an anticline. Computer modeling consistently shows that if there were a fault in such a position, it would deform the hangingwall section far beyond the immediate area of poor data. If the record shows no such disturbance, it is almost certain that there is no fault present. In general, modeling shows there is no justification for the common assumption that faults in the subsurface extend to the surface, steepen upward and die out in shales. Similarly, the very act of rigorously balancing cross-sections shows that balancing must be applied with care. The classic "compression box" model of fold and fault generation implies that the entire deformed sedimentary sequence has been deformed in the same way, by the same amount and at the same time. These are not valid assumptions, and computer modeling may require different structural levels to be treated independently.

Computer modeling, at this stage, does only what can be done laboriously with a pencil and a sheet of paper. Because of its speed and accuracy, however, the resulting interpretation is normally the best of a large number of alternatives that would not have been otherwise tested. Also, by isolating the geometric aspects of deformation, computer modeling may well provide insights into the mechanics and history of deformation and, just as importantly, into the associated migration and entrapment of petroleum. **LE**

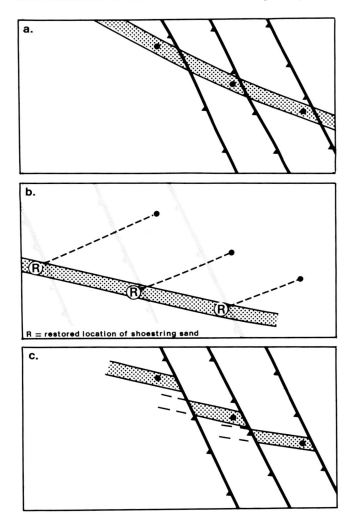

Figure 2. (a) Map showing apparent (false) trend of shoestring sand (stippled) through three oil wells in thrusted terranes. (b) True trend of sand shown by palinspastic map on which penetrations of the sand are restored to the locations where they were deposited before thrusting. Positions of thrusts before movement are shown by lightly stippled lines. (c) Present true trend of shoestring, derived from palinspastic map (b), combined with present positions of wells and thrust faults. (Note offsets of sand across thrust faults.)

Peter B. Jones was educated at universities in both the United Kingdom and the United States. He worked for various Canadian oil companies from 1955-80 before setting up International Tectonic Consultants Ltd. which specializes in petroleum exploration in deformed belts worldwide. Since 1981 he has been a partner in Thrustbelt Systems Ltd. where he provides geologic input in the development of the company's software for synthesizing balanced structural cross-sections (created by Helmut Linsser, a geophysical consultant and systems analyst). Author of more than 25 papers on structural geology, Jones has given lectures and courses to oil companies and universities in Europe, North America and Asia.

SEISMIC INTERPRETATION 11

Interactive interpretation of a submarine fan, offshore Ireland: A case history

By D. BRADFORD MACURDA, JR.
and
H. ROICE NELSON, JR.

During the Eocene, a series of submarine fans developed in the Porcupine basin off the southwest coast of Ireland. Interactive interpretation of one—the Clontarf fan—shows the successive development of four sequences. Seismic facies analysis of these sequences suggests a progressive evolution from a more distal to proximal environment during deposition of the fan. Seismic facies mapping reveals an extensive series of sand-prone distributary systems in the youngest fan sequence. Amplitude anomalies suggest that this and other fans in the basin are favorable stratigraphic and/or structural traps for hydrocarbons.

The Porcupine basin is an oval-shaped reentrant in the shelf margin southwest of Ireland (Figure 1). The basin measures approximately 300 km long in a north-south direction and 120 km wide in an east-west direction. Subsidence began in the lowermost Cretaceous, and rifting extended from the Berriasian through Hauterivian (see *Structure and development of Porcupine Seabight sedimentary basin, offshore southwest Ireland,* Masson and Giles, 1986, AAPG *Bulletin*). Continued subsidence and sedimentation subsequently resulted in a thick section that exceeds 7 km in the deepest portion of the basin. Cretaceous sedimentary basin fill is composed of a heterogeneous succession of environments and lithofacies. During the Cretaceous, volcanic intrusions were emplaced in the lower part of the sedimentary fill. Chalk was the principal lithofacies deposited in the Upper Cretaceous. In the Lower Tertiary, subsidence accelerated, resulting in a succession of deep-water siliciclastic environments. Present water depths in the area of investigation exceed 1000 m.

Floored by continental crust, the basin has a medial volcanic ridge oriented north-northwest to south-southeast. The eastern basin margin is oriented in the same direction. Masson and Giles inferred that Hercynian movements of the Late Paleozoic produced zones of weakness having a north-northwest to south-southeast orientation. A subsidiary fault trend oriented north-south to northeast-southwest is associated with the Permian through Jurassic history of the basin. During this time the primary tectonic setting was that of a shelfal succession on a passive margin. Rifting broke this succession apart in the Lower Cretaceous. The Porcupine basin is also marked by a broad north-south gravity low that mirrors bathymetric trends. Gravity modeling suggests the deep Porcupine basin was formed by subsidence of an area of thinned continental crust.

Wells in the northern part of the basin have penetrated shallow water sediments, including some coals. During the Paleocene and Eocene, an extensive siliciclastic depositional system prograded into Porcupine basin from the north, forming a thick section that thins dramatically from north to south (Figure 2). This differential infill, combined with continued subsidence, formed a pronounced bathymetric gradient between the margins of the basin and its center. During the Eocene, a series of submarine fans developed seaward of the Eocene shelf margin. In east-west seismic profiles, these appear as a series of distinct mounds. In plan view, each of the submarine fans has an elongate oval shape. Internal reflection character suggests that each submarine fan was sourced from a different point around the periphery of the basin. In the following section we will describe the seismic stratigraphy and internal development of one of these fans.

The Clontarf fan, located in the west central portion of the basin, was selected for detailed investigation. A grid of seven east-west and four north-south lines was available for us to investigate the seismic stratigraphy of this fan. Seismic data were acquired by Merlin Geophysical in 1981, using an energy source of 950 in^3 air guns and a cable length of 2920 m. The data were sampled at 4 ms. Processing included finite difference migration and a time-variable filter. The data were scaled using a short AGC operator. Data loaded into a Landmark workstation consisted of a subset of the final processed data—through migration. Spacing of the lines in both the north-south and east-west directions is 6.5 km.

The Clontarf fan is ovoid in shape with the long axis oriented northwest-southeast, tapering to the southeast. It is approximately 40 km long and 25 km wide at maximum extent. The extreme northwestern tip of the fan is unknown because it extends beyond the survey limits. The total thickness of the fan is about 600 m. In plan view, it is similar in shape to the Rhone fan and the Indus fan, but the Clontarf fan is much smaller and thinner. The top of the Eocene siliciclastics underlying the Clontarf fan are recognizable by toplap. Siliciclastics which infilled around the Clontarf fan show onlap and downlap in areas between the fan and those adjacent to it. Interpreted as fine-grained, deep-water sediments, these sediments provide a seal to the Clontarf fan.

The fan is divisible into four internal sequences. Evidence for this consists of the external geometry of each sequence, as well as internal downlap and onlap at the margins of the fan. We have labeled these as sequences A through D. Sequence boundaries are shown in Figure 3. The top of the Eocene prodelta is marked by the lowermost boundary, shown in red. Sequence A, the oldest in the Clontarf fan, is shown by the purple horizon. Geometrically, it is a sheet and extends across the entire basin. Internally, the purple sequence is characterized by parallel reflectors with little variation. The second sequence (B), shown by the green horizon, also has a sheet geometry but lacks the regional extent of the purple sequence. Internally, most of the reflectors are parallel, although some prograding clinoforms are found. The third sequence (C), marked by the orange horizon, is the oldest sequence that shows a distinctly mounded appearance in cross-section. Internally, seismic facies in the orange sequence are a combination of hummocky clinoforms and parallel reflectors, with shingled reflectors occurring toward the periphery of the fan. The youngest (D) is shown by the blue horizon. Externally, it has a very strong mounded expression. Internally, the seismic facies are a combination of hummocky clinoforms and parallel reflectors in the northwestern or proximal portion of the fan. The percentage of hummocky clinoforms within the sequence consistently increases downdip to the southeast. Shingled reflectors are found at the margins of the fan in the blue sequence.

Sequence boundaries were independently identified on two north-south and two east-west lines. Then these were correlated and extended throughout the entire grid of lines on which the fan appeared. Figure 4 is a fence diagram showing the position of the sequence boundaries on each of the lines analyzed. The vertical lines mark intersections of crossing lines of control. On the workstation, fence diagrams like these can be rotated and viewed from many different perspectives, allowing the analyst to appreciate the 3-D configuration of each sequence in the Clontarf fan. Computer-generated horizon maps of sequence boundaries can also be viewed in perspective, and it is possible to compare the attitude and geometry of several contoured surfaces in this way.

As seen in seismic sections, the geometry of the sequences has been somewhat distorted by additional subsidence that occurred after deposition. In order to approximate the original external geometries and to examine the progressive development of the Clontarf fan, horizon flattening can be used. The relative thickness of the four sequences becomes obvious when sequence A at the base is flattened on the top of the Eocene prodelta, sequence B is flattened on top of sequence A, and so on. When these flattened sequences are displayed in the same order that they were deposited (Figure 5), progressive change from the sheet geometry of sequence A to the mounded geometry of sequence D is evident.

Maps and cross-sections. One of the most utilitarian aspects of the workstation is the ability to create a series of maps that shows sequential development. First, an isochron contour map of the total fan thickness was generated to show the location and extent of the fan. Seismic control does not confirm closure in the northwestern part of the fan. Contours derived from this map were overlaid on maps of each sequence.

Figure 6, for example, shows three different aspects of sequence D. The left side displays the original 2-D isochron control as ribbons. The map in the center is a smoothed, gridded version of this control, with relative thicknesses highlighted by the color bar on the right. Creation of this gridded map from 2-D control took place in three stages. First, a triangulated surface was fitted across the 2-D control, which in this case represented an isochron map between the top of sequence C and sequence D. Second, a grid was defined, and the system wrote the interpolated isochron values into each grid point. Third, the linear interpolated map was smoothed. Overlying the sequence map is the contour map, in two-way traveltime, showing total thickness of the fan.

The isochron map of sequence A, the oldest portion of the fan, exhibits a maximum of only 15 ms variation in thickness. In contrast, the isochron map of sequence D shows that over 500 ms of section were deposited along the axis of the fan. As the fan developed seaward of the Eocene deltaic margin, maximum differentiation between the fan and the surrounding deeper portion of the basin occurred during deposition of sequence D. Thinning into troughs bordering the fan is clearly evident in Figure 6.

Interactive display allowed us to create a set of panels consisting of a series of cross-sections across the length and width of the fan. This kind of display facilitates the examination of downdip changes in geometry and seismic facies from the more proximal to distal portions of the fan.

Another effective method for understanding the deposition of the fan is to display different aspects of data for the same line. Figure 7 is a zoomed view of sequence D in an east-west cross-section through the western half of the fan. The top panel is a seismic section display, while instantaneous phase is displayed in the center and reflection strength on the bottom. Windows like these may be displayed simultaneously on the same screen using 8 bit pseudocolor. The phase display in the center shows much more lateral continuity than regular seismic amplitudes. The strength of the reflection interfaces at sequence boundaries C and D is seen best in the display at the bottom.

At shotpoint 205 in Figure 7, a slight bump at the top of sequence D is distinctly visible. After examining various attribute displays, we interpreted this feature as the result of differential compaction between a sand-prone channel system and shale-prone interchannel deposits. A similar feature can also be seen at the top of sequence C near shotpoint 450.

Seismic facies within the Clontarf fan exhibit gradational changes both vertically and horizontally. Sequence A, as noted earlier, is characterized internally by parallel reflectors. But the relative percentage of hummocky clinoforms within the fan increases upward, reaching a maximum in sequence D. Laterally, the relative percentage of hummocky clinoforms increases downdip from north-west to south-east in sequence D. Hummocky clinoforms are interpreted as the distributional axes of more sand-prone facies within the fan. The degree of bifurcation is interpreted to increase downdip, as seen in channel systems of many modern submarine fans. The distribution of hummocky clinoforms in the lower and upper half of sequence D was mapped separately and contoured to show the distribution of channel-prone facies. Figure 8 shows a schematic representation of the channel system in the lower half of sequence D overlaid on the contoured outline of the fan. We believe the channel systems reached a point of instability and shifted position during the development of sequence D, resulting in different channel configurations during deposition of the lower and upper halves of the sequence.

The purpose of seismic facies mapping is to provide insight into the sedimentological history of the sequence being mapped. Thus, if we proposed a well location to test the hydrocarbon potential of the Clontarf fan, we would combine stratigraphy and structure to choose an optimum location. When the two-way traveltime section is converted to depth, the slope on the top of the Eocene delta is flattened, becoming more nearly horizontal. This increases the presumed closure in the northern portion of the fan. A comparative overlay of the seismic facies and structure contour map suggests a local culmination in the northeastern portion of the Clontarf fan. We believe this constitutes a favorable prospect for testing the hydrocarbon potential of this submarine fan.

Before actually drilling the initial test, similar seismic strati-

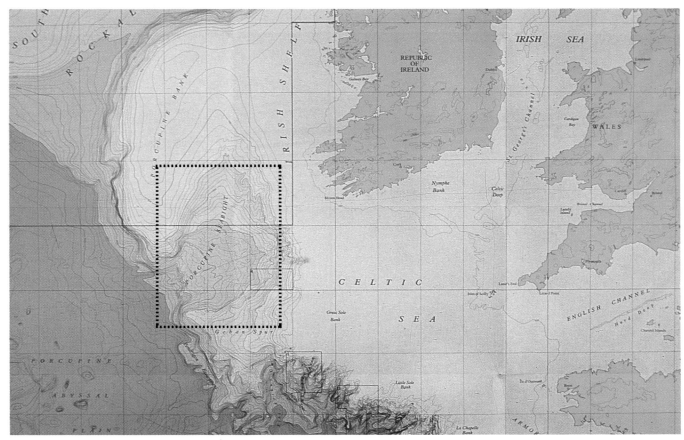

Figure 1. Location of the Porcupine basin.

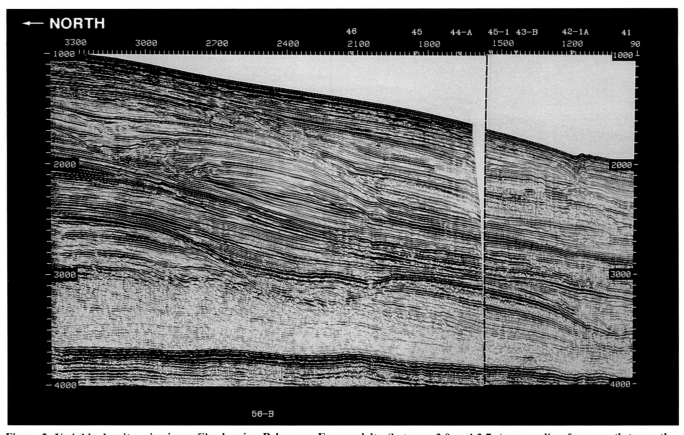

Figure 2. Variable density seismic profile showing Paleocene-Eocene delta (between 3.0 and 3.7 s) prograding from north to south.

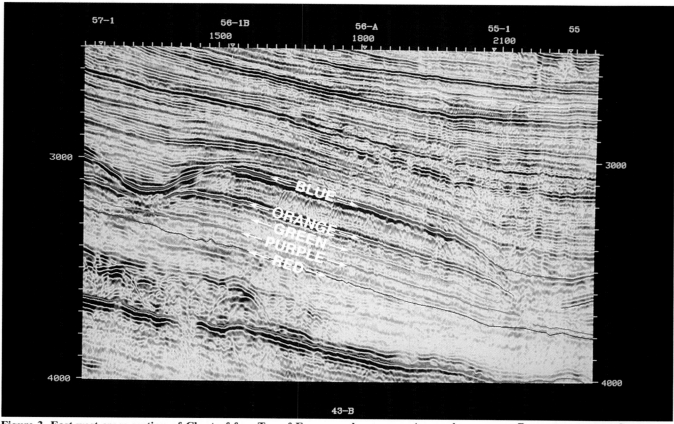

Figure 3. East-west cross-section of Clontarf fan. Top of Eocene, red; sequence A, purple; sequence B, green; sequence C, orange; sequence D, blue.

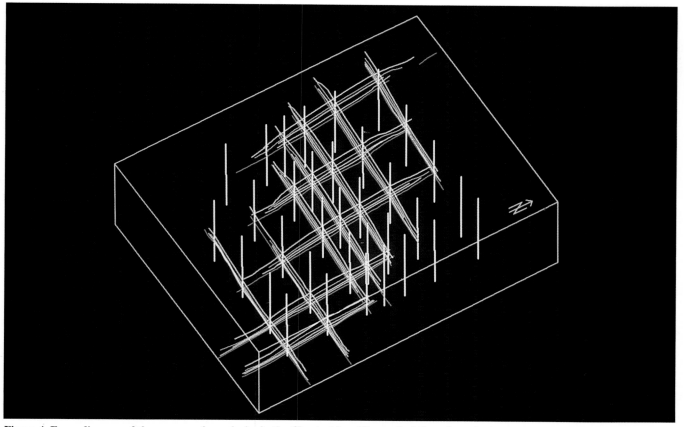

Figure 4. Fence diagram of the sequence boundaries in the Clontarf fan. The vertical lines show the intersections of crossing control lines.

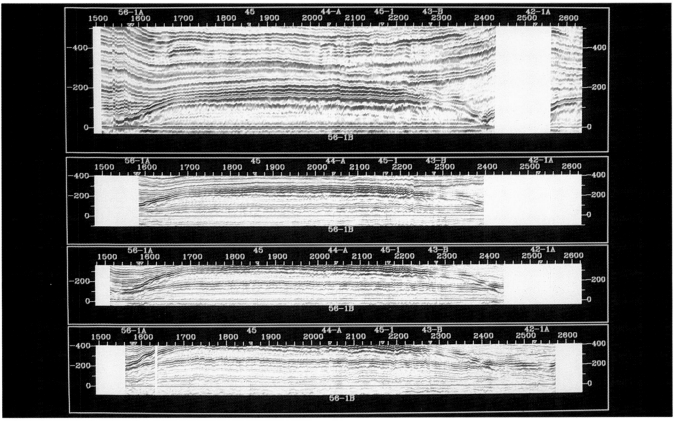

Figure 5. Four windows using horizon flattening to show relative thickness of sequences. D at top is flattened on C; A at base is flattened on Eocene prodelta. In each window, the top of that sequence is green; top of previous sequence is purple.

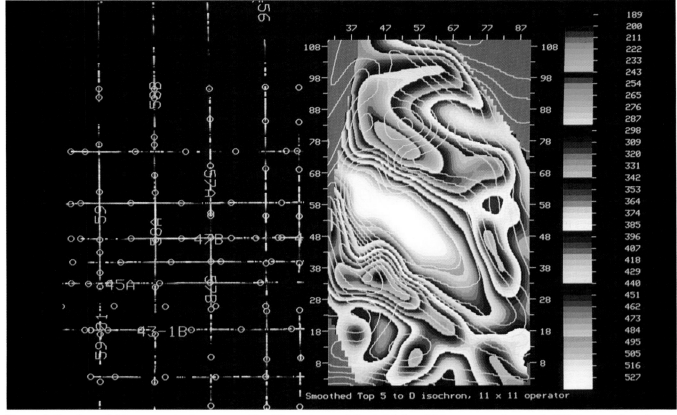

Figure 6. Isochron thickness map of sequence D with contours of the total fan thickness superimposed. Thickest section shown by grays/white along axis; thinnest sections shown by blue-greens in troughs bordering axis.

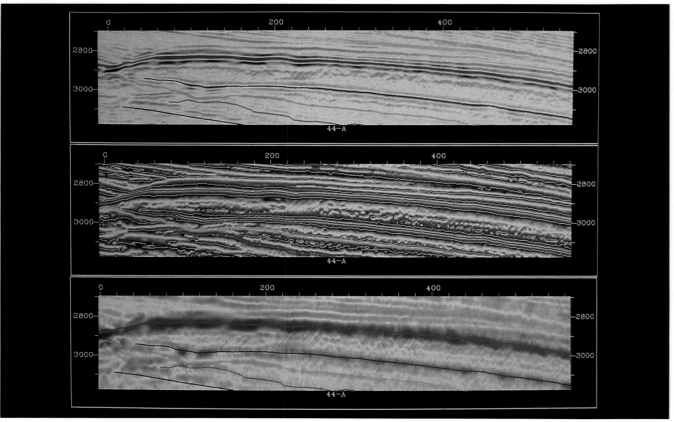

Figure 7. Different attribute displays of a single east-west line, with color-coded sequence boundaries. Top window shows a colored seismic section display; center, instantaneous phase; bottom, reflection strength.

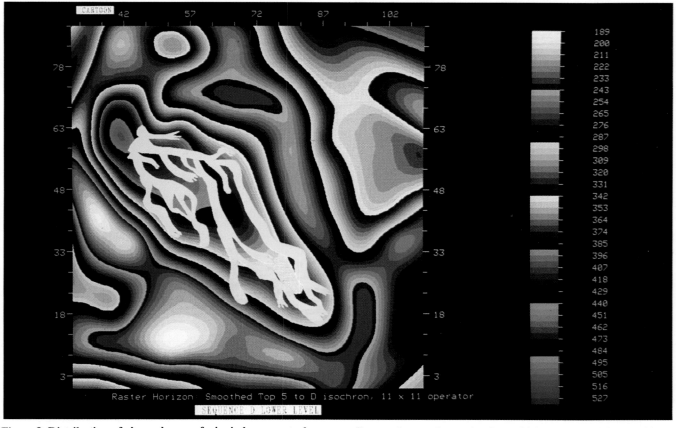

Figure 8. Distribution of channel-prone facies in lower part of sequence D, superimposed upon isochron thickness map of Clontarf fan.

graphic and facies analyses should be conducted for other submarine fans at this stratigraphic level, so the most promising fan could be selected. Two amplitude anomalies encourage us to believe hydrocarbons are reservoired in the submarine fans of Porcupine basin. The first anomaly is a flatspot at a local culmination on a fan located 40 km east of the Clontarf fan. The second is a gas chimney that occurs above the southeastern portion of the fan. Comparative facies analysis of overlying sequences suggest that the most probable source is sequence D in the submarine fan.

In conclusion, initial Lower Tertiary deposition in the Porcupine basin was dominated by the development of a deltaic succession sourced from the northern portion of the basin. Increased subsidence during this interval of time resulted in increased bathymetric differentiation between the margins and center of the basin. During the latter part of the Eocene, a series of submarine fans developed seaward of the Eocene shelf margin, sourced from different points around the periphery of the basin. The Clontarf fan developed in the western portion of the basin. Internally, the fan is comprised of four sequences, each of which developed in a more proximal setting than its precursor. From seismic sequence and facies analysis, environments are interpreted as a series of distributary systems that bifurcated in a southeasterly direction. Local highs within the fan sequence probably resulted from differential compaction between sand-prone and shale-prone lithofacies. Extensive development of sand-prone facies suggests that, under favorable economic conditions, the Clontarf fan will become a viable prospect.

Present water depths over the fan make it a rank wildcat. But recent discoveries in the southern hemisphere of submarine fans containing three billion barrels of reserves each should encourage exploration for sands in other basins also deposited in deepwater environments. **LE**

Acknowledgments: We would like to thank both Merlin Geophysical for making available the data utilized in this investigation, and Landmark Graphics Corporation for providing workstation time. This paper was originally presented at the 49th EAEG meeting in Belgrade, Yugoslavia, in June 1987 and later, in October, at the 57th SEG Annual International Meeting in New Orleans.

D. Bradford Macurda, Jr., received a B.S. (1956) and a Ph.D. (1963) in geology from the University of Wisconsin. From 1963-78, he was a professor at the University of Michigan. In 1978 he went to work for Exxon Production Research, and in 1981 joined The Energists in Houston, Texas, where he is presently serving as vice-president. His professional memberships include the SEG, EAEG, AAPG, AGU, GSA, AAAS and SEPM. Macurda's main interests are stratigraphy, geophysics and sedimentology.

H. Roice Nelson, Jr., earned a B.S. in geophysics from the University of Utah and an M.B.A. from Southern Methodist University. He worked for Mobil Exploration and Producing Services from 1974-78, was general manager, AGL, for the University of Houston, and in 1982 helped found Landmark Graphics Corporation where he currently serves as senior vice-president. Nelson is a member of SEG, EAEG, AAPG, GSH and the Norwegian Petroleum Society. He is the author of New Technologies in Exploration Geophysics.

Wavelet estimation: An interpretive approach

By RAYMON L. BROWN, WENDY McELHATTAN and DONALD J. SANTIAGO
University of Oklahoma and Oklahoma Geological Survey
Norman, Oklahoma

There appears to be no uniform agreement on how to determine the polarity and shape of the wavelet in seismic data. In addition, when the determination is performed, the interpreter is usually the last to be informed of the results. Although the wavelet estimation problem can become buried in mathematics, the interpreter is best suited to judge the quality of the estimated wavelet and the applicability of the wavelet to the final interpretation. An approach to getting the interpreter involved in the wavelet estimation process is described in this article, and we believe that it offers some conceptual advantages over many of the popular methods currently in use.

Fortunately, structural interpretation can usually be performed without a detailed knowledge of the wavelet. However, in order to approach stratigraphic problems, an accurate knowledge of the wavelet is necessary. Once the wavelet is known, wavelet shaping or synthetic impedance logs can be used to simplify interpretations (e.g., the thickness of gas sands can be estimated or subtle stratigraphy can be mapped).

Once the wavelet is known, an interpreter has a powerful tool for the determination of the relative impedance of a reflector (i.e., the sign of the reflection coefficient may be determined). Such knowledge may be used in some exploration areas to evaluate "bright spots" without the need to consider amplitude-with-offset or V_p/V_s methods. Wavelet estimation is critical because an accurate estimate of the wavelet can prevent a misinterpretation of stratigraphy and can be used to determine the polarity of the reflections.

Wavelet estimation is difficult because the recorded seismic trace contains the wavelet convolved with the unknown earth response (reflectivity). Any noise that might be present further complicates the problem. All of these signals are affected by multiples, instrument response, source-receiver array responses, near-surface effects, attenuation and dispersion. Thus the wavelet is mixed with other signals which must be eliminated in order to obtain a measure of the wavelet. Since the seismic noise and earth response are usually unknown, the wavelet estimation process is not trivial.

A direct approach may be taken by measuring the wavelet when it is effectively free of the earth response. For marine data, the source can be taken to the deeper part of the ocean where the downgoing wavelet can be measured directly by a hydrophone towed at great depths. However, these measurements then must be augmented by including the effects of the receiver ghost reflection that are present when the receiving hydrophones are near the marine surface. These direct methods usually do not account for the radiation pattern of the source. In addition, the effects of stacking and migration are not included in this type of "deterministic estimation" method. On land, wavelets can be measured down the borehole (e.g., VSPs). Here, the near-surface and free-surface ef-

Of the techniques available on the open market for wavelet estimation, few allow the interpreter any input to the process until the final wavelet is obtained.

fects must be taken into account as well as the radiation pattern of the source. Surface-consistent estimation methods attempt to remedy some of these near-surface problems.

As a result, "deterministic" wavelet methods must make a number of estimates which leave some questions regarding the final wave shape left on the data. Since the final version of the processed line has all of the processing and radiation effects incorporated into the wavelet, estimation directly from the seismic data eliminates the necessity of having to calculate the effects not included in a direct measurement. The wavelet estimation can be accomplished with only the seismic data or with the seismic data and available well logs. The well logs add to the determination of a wavelet a constraint that reduces the number of assumptions necessary.

Of the techniques available on the open market for wavelet estimation, few allow the interpreter any input to the process until the final wavelet is obtained. An interactive approach to wavelet estimation might be desirable for the situation in which both seismic and well log data are available. The value of this approach is that the interpreter can repetitively evaluate the estimated wavelet by comparing the predictions of the wavelet to the known impedance contrasts within the well (or vice versa).

Much of what can be accomplished interactively will be based upon methods already in existence. For this reason, a brief review of selected wavelet estimation methods is offered, along with a description of some of the basic assumptions.

Wavelet estimation from seismic data only. A classical approach to wavelet estimation directly from the processed seismic data involves making the model assumption that the data are the result of convolving a minimum phase wavelet (which does not change shape along the section) with a random series of reflectors. This statistical approach to wavelet estimation began with the work of the Geophysical Analysis Group at MIT in the early 1950s, and many of the basic ideas derived at that time are still in use today. More specifically, in the statistical method of wavelet estimation, the impulse response of the earth is assumed to be white, random and stationary. All of these assumptions can be invalid in practice, but the applicability of the statistical method rests upon the size of the error rather than the question of whether an error exists due to a basic assumption.

Theoretically, the amplitude spectrum of the wavelet for the model described above can be obtained directly from the processed seismic trace. However, the assumption of the random nature of the earth's reflectors often fails.

Under these conditions, remedial methods are applied. These include (a) averaging over many time windows to insure a random reflectivity and (b) using well log estimated reflectivities to pre-whiten the spectrum of the trace. Once the spectrum is obtained, the minimum phase wavelet may be extracted by a number of methods.

However, there are many circumstances where the minimum phase assumption causes problems. For both land and marine environments, the seismic wavelet can be significantly different from the assumed minimum phase wavelet. Turhan Taner developed an approach to wavelet estimation which did not require the minimum phase assumption. He suggested that a filter be applied to the data that forces the wavelet in the data to be minimum phase. Next, he applied the statistical methods described above to estimate the forced minimum phase wavelet. Finally, he backed out the filter which forced the wavelet to minimum phase in order to obtain the original mixed phase wavelet. The danger of this method is that (if abused) it can change the reflectivity and ultimately the interpretation.

A better understanding of Taner's approach to wavelet estimation may be obtained by considering the mixed phase wavelet in terms of the z transform. (A brief description of the z transform is given in *Measuring wavelet phase from seismic data* by J.P. Lindsey, July 1988 *TLE*. Lindsey presents a method for phase estimation that is based upon the movement of roots in the complex plane—a concept that is also the foundation of Taner's approach.) Assume the wavelet has been expressed as a sequence of numbers which represent samples taken at a constant interval in the time domain. This series may be expressed as a polynomial via the z transform; this polynomial is a summation but it can be rewritten as a product of terms containing the roots (values which make the expression equal to zero). If all the roots have an absolute value greater than one, the associated wavelet can be shown to be minimum phase (i.e., most of the energy arrives in the early portion of the wavelet).

Figure 1 shows an example wavelet in the time domain and its roots plotted in the z-transform domain. Since all of the roots are outside of the circle with radius of unity, the example wavelet is a minimum phase wavelet. If some of the roots have absolute value less than one, the wavelet is mixed phase. If all the roots have absolute value less than one, the wavelet is maximum phase.

Taner's method is based upon an idea described by R.W. Schafer of MIT in 1969. Schafer pointed out that when the z transform of a wavelet is multiplied by an exponential sequence, the roots of the wavelet change position. So the central idea of Taner's method is to multiply the z transform by something which will make each root of the resulting polynomial have an absolute value greater

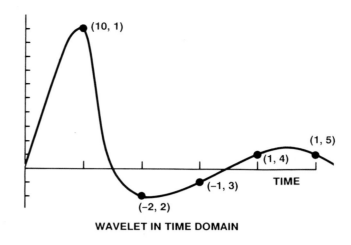

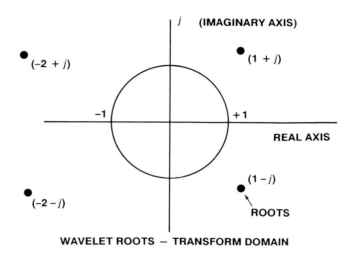

Figure 1. An example wavelet in the time domain and its roots plotted in the z-transform domain.

than unity; this insures that the resultant wavelet will be minimum phase and the statistical methods described earlier can be used.

However, the theory weakens a little at this point, because there appears to be no *a priori* method of selecting the optimum value (which should be just less than one) of the exponential factor which will be used as the multiplier. In practice, we have a more complicated problem since the wavelet is mixed with other signals, and we are in the dangerous position of being able to modify the reflectivity (and ultimately the interpretation of the data) as well as the wavelet. Taner has suggested an empirical approach in which a value just less than one (e.g., 0.98) is tried. Then a statistical estimate of the wavelet is made. Next the process is repeated with a smaller factor and another wavelet estimate is made. If the second estimate is significantly different from the first, the value of the factor must be decreased until the wavelet estimate between two successive values stabilizes. Once the stable estimate of the forced minimum phase wavelet is obtained, the exponential filter is "backed out" to obtain the estimated mixed phase wavelet.

Another method of wavelet estimation—homomorphic deconvolution—runs into similar problems when attempting to separate the wavelet from the reflectivity in what is called the "complex cepstrum domain" (which is essentially an inverse Fourier transform of the logarithm of the Fourier transform of the seismic trace). This method will not be discussed further because it has not proved a popular choice in the industry.

It should be stressed that mixed phase wavelet estimation methods, Taner's method and homomorphic deconvolution can change the reflectivity and, therefore, the interpretation. The skill and experience of the processor are critical when these techniques are applied.

Wavelet estimation from well logs. When well logs are available, we should be able to compute an impedance profile which corresponds to the seismic traces near the well control. The impedance log can then be used to estimate the reflectivity sequence. Once the reflectivity sequence is known, the wavelet may be estimated using a number of methods.

The prediction of the reflectivity sequence associated with the seismic data is not a trivial step. In our opinion, the accurate estimation of the reflectivity sequence from the well log data can be the hardest part of the problem when attempting to use seismic and well log data to make an estimate of the wavelet. This is discussed in further detail in the section on interactive wavelet interpretation.

Assuming that the reflectivity sequence can easily be predicted from the well logs, at least three approaches to the problem of wavelet estimation are possible.

The first method—spectral division—involves simply dividing

The prediction of the reflectivity sequence associated with the seismic data is not a trivial step.

---

the Fourier transform of the seismic trace by the Fourier transform of the reflectivity sequence and then inverse Fourier transforming to obtain the estimated wavelet.

The second method—Wiener least squares filtering—estimates the wavelet in the time domain using the Wiener filter and the known reflectivity sequence.

A third method involves geophysical inversion in which the wavelet is the unknown to be determined from the known reflectivity and seismic trace. The inversion scheme may be set up by forward modeling to compute a synthetic seismogram. The synthetic seismogram is then compared to the seismic trace. The wavelet is varied until the synthetic gives a satisfactory match to the observed data. Since it is not practical to search through all possible combinations of wavelets which might produce the desired synthetic seismogram, various approaches have been designed to aid the search for the unknown parameters.

A problem that occurs with spectral division is the possible division by zero (notches in the reflectivity spectrum). It is customary to whiten (i.e., add a constant to the Fourier transform) of the reflectivity sequence to prevent this from happening.

In the time domain, the Wiener least squares estimate is considered superior to spectral division because it gives a more reliable estimate in the presence of noise.

Inverse methods offer a powerful tool for the simultaneous estimation of the wavelet and impedance (or reflectivity sequence). In addition, inverse methods yield a measure of how sensitive the observed data is to the model parameters.

One approach to this type of inverse method was described by D.A. Cooke and W.A. Schneider in *Generalized inversion of reflection seismic data* (GEOPHYSICS, 1983). They minimized the number of parameters by describing the wavelet amplitude spectrum in terms of a trapezoid defined by four frequencies. The phase was assumed to be a simple linear phase shift. The impedance in each interval was described by the impedance at the top of the layer, the gradient of the impedance, and the two-way traveltime through the interval.

Once the forward model parameters are assigned and an initial guess is made, a synthetic trace can be computed. Next, a correction to the initial guess model is computed using the differences between the synthetic and the observed data. The computed correction to the initial model improves the fit to the observed data. Then a synthetic for the first improved model is computed. If the errors are still considered large, the process is repeated until the desired fit is obtained.

If the reflectivity is assumed to be known from well logs, it has been shown (by Larry Lines and Sven Treitel in *Wavelets, well logs and Wiener filters*, 1985 *First Break*,) that the general linearized inverse is equivalent to the Wiener least squares approach.

Synthetic seismograms and reflectivities. At this point we question not only the process of estimating reflectivity from the logs, but the general procedure of making synthetic seismograms that match the observed data.

If the log and reflectivity are complicated, the successful match with the data is not, in general, a great aid to the interpretation if stratigraphy is the target of the study. The interpreter needs to be able to decompose the seismic data so that the reflections from specific horizons may be identified. A brute force approach to this problem consists of plotting the isolated wavelets from each reflector displaced horizontally adjacent to the composite trace so that individual reflectors can be identified. Inverse methods can be used to simplify this process by using the known wavelet to estimate a synthetic impedance log.

However, there is no easy method for estimating the wavelet which can give the interpreter confidence in the data in order to make an interpretation. Equally as difficult, but discussed far less in the literature, are the problems faced when attempting to estimate the reflectivity from the log data. Logs have been blocked, filtered, squeezed or stretched in order to get agreement with the observed data. The fact that logs are treated this way indicates that prediction of the seismic reflectivity from well logs is as much a problem as is the wavelet estimation. The difficulty is not all that surprising since the model typically used for log data is a plane wave incident upon a horizontally layered media.

Typical reasons for a poor fit of synthetic to observed data are:

- Wrong wavelet used in synthetic.
- Neglect of multiples.
- Inaccurate time-to-depth information.
- Inaccurate log measurements.
- Anisotropy.
- Differences between plane wave incident on layered media and actual data collection and processing.

These reasons and others can be used to explain the poor fit of synthetic data to observed seismic data. Thus we wonder just how the log can be used if there are so many variables which prevent a simple prediction of reflectivity. To address these problems, we recommend selecting portions of the log which have simple (unambiguous) interpretations and which are easily used to estimate the reflectivity.

For example, high-amplitude continuous reflections due to a reflector which is uniform over a region might be easily associated with a known lithology in the well control. These high amplitude reflections are often associated with simple portions of the log which are used to estimate the reflectivity for this portion of the log (with or without multiples).

By restricting the wavelet estimation process to only the selected window about a particular high-amplitude reflection, the seismic signal will be expected to have a higher signal-to-noise ratio. Thus, we suggest avoiding the use of all of the seismic section at once since any estimate based upon the complete seismic trace must account for the seismic noise. If the chosen window in time on the seismic data corresponds in depth to simple portions

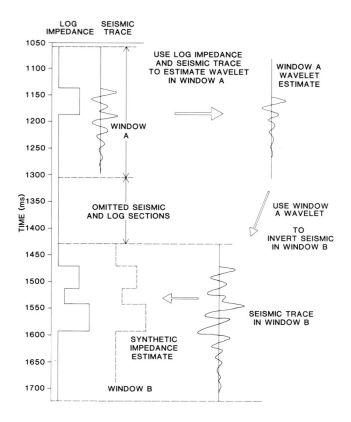

Figure 2. Interactive scheme in which the wavelet is estimated in one window of the seismic/log data and is inverted in a second window.

of the well log where the reflectivity is easily predicted (e.g., a single thick interval easily identified on the seismic section), then a double advantage is gained over conventional methods which attempt to solve the complete problem all at once.

If nothing else is attempted, the ideas presented above could be used on selected windows using any one of the three methods described for estimating the wavelet from well logs. However, we can go further by using an interactive scheme in which the wavelet is estimated in one portion (window) of the seismic/log data and is inverted or modeled in a second window. For example, the wavelet estimate obtained from the first window may be used to invert the seismic trace to a synthetic impedance log in the second window and then compared to the actual impedance log calculated from the well control (see Figure 2.)

An alternate approach could consist of generating a synthetic from the impedance in the second window and comparing it to the observed seismic data. The impedance contrasts associated with high-amplitude reflectors are often quite simple. For example, a thick limestone extending over a region may have a simple impedance contrast which can be used in one window. Within the window, the interpreter needs the ability to edit the sonic and density logs used to estimate the reflectivity. In addition, some variability of the assumptions used to compute the reflectivity is desirable (e.g., with and without multiples and attenuation). A choice of wavelet estimation methods might also be desirable.

Thus, high-amplitude reflections from marker beds are used to make a wavelet estimate for a particular time-depth window on the seismic/log data. Once the wavelet has been estimated for one window, we use that estimate to invert a second window. If the computed seismic impedance agrees with the prediction from the logs, the estimated wavelet is then used to predict changes away from the well control within the time-depth domain of the windows. If the agreement is close but not satisfactory, some effort to account for the extra attenuation and transmission effects between the windows may be necessary. In addition, the log data may need further editing to improve the fit. Then the interpreter can compare the wavelet estimate from the first window using two different assumptions for the reflectivity (e.g., with and without multiples or with and without editing).

The advantage of this process is that the interpreter can interactively play a part in the wavelet interpretation process. Admittedly, interpreters are frequently not very strong in signal processing. However, geologic insight has for years been the final judgment on processing quality.

The system proposed here would allow the interpreter to make the decision, based upon his knowledge of the geology, on whether the estimated wavelet is credible. In addition, a technical advantage is gained by selecting high-amplitude, simple geometry reflectors. When attempting to estimate reflectivity from the complete well log, we can introduce a great deal of uncertainty into the process of wavelet estimation. Thus, elegant mathematical approaches can become limited by the quality of the input. The interpreter with an interactive system can get around these difficulties.

Of course, all that glitters is not the wavelet. Within our selected window, we frequently have other signals to be eliminated from our estimation process. However, any of the methods described in this article could be changed to account, first, for large reflectors within the window and, then, for second order effects in and around the selected window.

Wavelet estimation methods incorporate many assumptions. We eliminate the need for many of them by using sonic and density logs. However, prediction of the seismic reflectivity sequence from well logs is not an easy task. It requires a number of decisions regarding the assumptions and editing processes.

Automated methods of wavelet estimation suffer from the diversity of situations that occur and are subject to failure without notice. Admittedly, there is a lot of publicity about making machines into intelligent decision makers. However, by involving the interpreter interactively with the process, we feel that many high-powered wavelet estimation methods can be beaten due to their intrinsic reliance on the quality of the input data. **L̲E̲**

Raymon L. Brown received a B.S. in physics (1967) from the University of Texas at Arlington, an M.S. in physics (1969) from the University of Hawaii, and a Ph.D. in geophysics (1975) from MIT. He has worked for Shell Development and Exxon Production Research, and also for Exxon USA, Kerr-McGee, and a reservoir engineering company (Miller and Lents, Ltd). Brown is currently a visiting associate professor of geophysics at the University of Oklahoma in a joint appointment with the Oklahoma Geological Survey and the university's School of Geology and Geophysics.

Wendy McElhattan is a graduate student in geophysics at the University of Oklahoma. She received a B.S. in earth science/geology from Clarion University of Pennsylvania. She is a junior member of AAPG and a student member of SEG and the Geophysical Society of Oklahoma City. McElhattan's current interests include the development of techniques for the determination of lithology from seismic data.

Donald J. Santiago received a B.S. in geology from Michigan State University and an M.S. in geophysics from the University of Oklahoma. From 1979-81 he worked as an interpreter for Exxon in the Oklahoma City office. Santiago joined GHK in 1981 as a staff geophysicist and, in 1983, went to work for Southwestern Energy where he is currently a senior staff geophysicist and geophysical department manager.